Lecture Notes in Geoinformation and Cartography

Series Editors: William Cartwright, Georg Gartner, Liqiu Meng, Michael P. Peterson

Tijs Neutens · Philippe De Maeyer
Editors

Developments in 3D Geo-Information Sciences

Springer

Editors

Tijs Neutens
University of Ghent
Dept. Geography
Krijgslaan 281
9000 Gent
Belgium
Tijs.Neutens@ugent.be

Dr. Philippe De Maeyer
University of Ghent
Dept. Geography
Krijgslaan 281
9000 Gent
Belgium
Philippe.Demaeyer@ugent.be

ISSN 1863-2246 e-ISSN 1863-2351
ISBN 978-3-642-26109-1 e-ISBN 978-3-642-04791-6
DOI 10.1007/978-3-642-04791-6
Springer Heidelberg Dordrecht London New York

Printed on acid-free paper

Springer is part of Springer Science+Business Media (www.springer.com)

Preface

Realistically representing our three-dimensional world has been the subject of many (philosophical) discussions since ancient times. While the recognition of the globular shape of the Earth goes back to Pythagoras' statements of the sixth century B.C., the two-dimensional, circular depiction of the Earth's surface has remained prevailing and also dominated the art of painting until the late Middle Ages. Given the immature technological means, objects on the Earth's surface were often represented in academic and technical disciplines by two-dimensional cross-sections oriented along combinations of three mutually perpendicular directions. As soon as computer science evolved, scientists have steadily been improving the three-dimensional representation of the Earth and developed techniques to analyze the many natural processes and phenomena taking part on its surface. Both computer aided design (CAD) and geographical information systems (GIS) have been developed in parallel during the last three decades. While the former concentrates more on the detailed design of geometric models of object shapes, the latter emphasizes the topological relationships between geographical objects and analysis of spatial patterns. Nonetheless, this distinction has become increasingly blurred and both approaches have been integrated into commercial software packages.

In recent years, an active line of inquiry has emerged along the junctures of CAD and GIS, viz. 3D geoinformation science. Studies along this line have recently made significant inroads in terms of 3D modeling and data acquisition. Complex geometries and associated topological models have been devised to approximate three-dimensional reality including voxels, polyhedrons, constructive solid geometry (CSG), boundary representation (B-rep) and tetrahedral networks. As input for these models, new technologies to collect three-dimensional data have become fully operational such as mobile mapping and 3D laserscanning. However, in light of these advances, up until now there is still a pressing need for robust 3D analysis and simulation tools that can be applied effectively in a wide range of fields such as urban planning, archaeology, landscape architecture, cartography, risk management etc.

In response to the lingering demand for 3D analysis and simulation tools, a workshop on 3D geoinformation was held in Ghent, Belgium on November 4-5, 2009. Following the successful series of past workshops, the Fourth International Workshop on 3D Geoinformation offers an international forum to promote high-quality research, discuss the latest developments and stimulate the dialogue between academics and practitioners with respect to 3D geoinformation, acquisition, modeling, analysis, management, visualization and technology.

This book contains a selection of full-papers that were presented at the workshop. The selection was based on extensive peer-review by members of the Program Committee. Only the most significant and timely contributions are included in this book. Selected contributors were asked to submit a revised version of their paper based on the reviewers' comments. All other papers and extended abstracts that were selected for oral or poster presentation at the workshop are published in a separate proceedings book.

The editors of this book would like to thank the many people who helped making this year's 3D GeoInfo workshop a success. We owe special thanks to Marijke De Ryck, Dominique Godfroid and Helga Vermeulen for their great help in organizing the conference, and Bart De Wit and Lander Bral for their excellent technological support. Thanks also go to Sisi Zlatanova for sharing experiences and advice on various aspects regarding the workshop, Agata Oelschlaeger for guiding us through the publication process and our sponsors for financial support. Finally, we would like to thank the members of the Program Committee for carefully reviewing the full papers and all those who submitted their work and participated in 3D GeoInfo 2009.

Ghent, Belgium
August 2009

Tijs Neutens
Philippe De Maeyer

Program Co-chairs

Programme co-chair Philippe De Maeyer
Ghent University (Belgium)

Programme co-chair Tijs Neutens
Ghent University (Belgium)

Local Committee

Marijke De Ryck, Dominique Godfroid, Helga Vermeulen
Ghent University (Belgium)

Program Committee

Alias Abdul-Rahman, University of Technology Malaysia (Malaysia)
Roland Billen, University of Liege (Belgium)
Lars Bodum, Aalborg University (Denmark)
Peter Bogaert, Geo-Invent (Belgium)
Arnold Bregt, Wageningen University and Research Centre (The Netherlands)
Volker Coors, University of Applied Sciences Stuttgart (Germany)
Klaas Jan De Kraker, TNO (The Netherlands)
Alain De Wulf, Ghent University (Belgium)
Claire Ellul, University college London (United Kingdom)
Robert Fencik, Slovak University of Technology (Slovakia)
Andrew Frank, TU Wien (Austria)
Georg Gartner, TU Wien (Austria)
Christopher Gold, University of Glamorgan (United Kingdom)
Muki Haklay, University College London (United Kingdom)
Thomas Kolbe, Technical University Berlin (Germany)
Jan-Menno Kraak, ITC (The Netherlands)
Mei-Po Kwan, Ohio State University (USA)
Hugo Ledoux, Delft University of Technology (The Netherlands)
Jiyeong Lee, University of Seoul, (South Korea)
Ki-Joune Li, Pusan National University (South Korea)
Twan Maintz, Utrecht University (The Netherlands)

Mario Matthys, University College Science and Art (Belgium)
Martien Molenaar, ITC Enschede (The Netherlands)
Stephan Nebiker, Fachhochschule Nordwestschweiz (Switzerland)
András Osskó, FIG/Budapest Land Office (Hungary)
Norbert Pfeifer, TU Wien (Austria)
Carl Reed, Open Geospatial Consortium (USA)
Massimo Rumor, University of Padova (Italy)
Mario Santana, K.U. Leuven (Belgium)
Aidan Slingsby, City University London (United Kingdom)
Uwe Stilla, Technical University of Munich (Germany)
Jantien Stoter, ITC Enschede (The Netherlands)
Rod Thompson, Queensland Government (Australia)
Marc Van Kreveld, Utrecht University (The Netherlands)
Peter Van Oosterom, Delft University of Technology (The Netherlands)
Nico Van de Weghe, Ghent University (Belgium)
George Vosselman, ITC Enschede (The Netherlands)
Peter Widmayer, ETH Zürich (Switzerland)
Peter Woodsford, 1Spatial and Snowflake Software (United Kingdom)
Alexander Zipf, University of Applied Sciences FH Mainz (Germany)
Sisi Zlatanova, Delft University of Technology (The Netherlands)

Contents

Contributing Authors

Behnam Alizadehashrafi
Dept. of Geoinformatics, Faculty of Geoinformation Science & Engineering, Universiti Teknologi Malaysia, 81310 Skudai, Johor, Malaysia

Pawel Boguslawski
Department of Computing and Mathematics, University of Glamorgan, Wales, UK

Katharina Bruhm
Dresden University of Technology, Germany

Manfred Buchroithner
Dresden University of Technology, Germany

Tet-Khuan Chen
Dept. of Geoinformatics, Faculty of Geoinformation Science & Engineering, Universiti Teknologi Malaysia, 81310 Skudai, Johor, Malaysia

Umit Isikdag

Eric Janssens-Coron
Geomatics Department, Pavillon Louis-Jacques Casault, Université Laval. G1K 7P4, Quebec, QC, Canada

Philippe De Maeyer
Department of Geography, Krijgslaan 281 S8, 9000 Ghent, Belgium

Jürgen Döllner
Hasso-Plattner-Institute at the University of Potsdam, Germany

Christopher Gold
Department of Computing and Mathematics, University of Glamorgan, Wales, UK
Department of Geo-informatics, Universiti Teknologi, Malaysia (UTM)

Benjamin Hagedorn
Hasso-Plattner-Institute at the University of Potsdam, Germany

Peter Hecker
Institut für Flugführung, Technische Universität Braunschweig.

Bernd Hetze
Dresden University of Technology, Germany

Dieter Hildebrandt
Hasso-Plattner-Institute at the University of Potsdam, Germany

Sudarshan Karki
Department of Environment and Resource Management, Queensland, Australia, University of Southern Queensland, Australia

Yohei Kurata
SFB/TR8 Spatial Cognition, Universität Bremen
Postfach 330 440, 28334 Bremen, Germany

Marc-Oliver Löwner
Institut für Geodäsie und Photogrammetrie, Technische Universität Braunschweig.

Kevin McDougall
University of Southern Queensland, Australia

Bernard Moulin
Computer Sciences and Software Engineering Department. Pavillon Adrien-Pouliot, Université Laval. G1K 7P4, Quebec, QC, Canada

Ivin Amri Musliman
Dept. of Geoinformatics, Faculty of Geoinformation Science & Engineering, Universiti Teknologi Malaysia, 81310 Skudai, Johor, Malaysia

Tijs Neutens
Department of Geography, Krijgslaan 281 S8, 9000 Ghent, Belgium

Kristien Ooms
Department of Geography, Krijgslaan 281 S8, 9000 Ghent, Belgium

Jacynthe Pouliot
Geomatics Department, Pavillon Louis-Jacques Casault, Université Laval. G1K 7P4, Quebec, QC, Canada

Alias Abdul-Rahman
Dept. of Geoinformatics, Faculty of Geoinformation Science & Engineering, Universiti Teknologi Malaysia, 81310 Skudai, Johor, Malaysia

Alfonso Rivera
Geological Survey of Canada. 490, rue de la Couronne. G1K 9A9, Quebec, QC, Canada

Andreas Sasse
Institut für Flugführung, Technische Universität Braunschweig.

Rod Thompson
Department of Environment and Resource Management, Queensland, Australia,
TU Delft, the Netherlands

Muhamad Uznir Ujang
Department of Geoinformatics,
Faculty of Geoinformation Science and Engineering,
University Technology Malaysia,
81310 UTM Skudai,
Johor Bahru, Malaysi

Izham Mohamad Yusoff
Department of Geoinformatics,
Faculty of Geoinformation Science and Engineering,
University Technology Malaysia,
81310 UTM Skudai,
Johor Bahru, Malaysi

Sisi Zlatanova

Euler Operators and Navigation of Multi-shell Building Models

Pawel Boguslawski[1] and Christopher Gold[2]

[1]Department of Computing and Mathematics, University of Glamorgan, Wales, UK
e-mail: pboguslawski@gmail.com
[2]Department of Computing and Mathematics, University of Glamorgan, Wales, UK Department of Geoinformatics, Universiti Teknologi, Malaysia (UTM)
e-mail: chris.gold@gmail.com

Abstract. This work presents the Dual Half Edge (DHE) structure and the associated construction methods for 3D models. Three different concepts are developed and described with particular reference to the Euler operators. All of them allow for simultaneous maintenance of both the primal and dual graphs. They can be used to build cell complexes in 2D or 3D. They are general, and different cell shapes such as building interiors are possible. All cells are topologically connected and may be navigated directly with pointers. Our ideas may be used when maintenance of the dual structure is desired, for example for path planning, and the efficiency of computation or dynamic change of the structure is essential.

Keywords: 3D Data Models, 3D Data Structures, Building Interior Models, Emergency Response, Disaster Management, Topology, CAD, Quad-Edge, 3D Dual Graph, 3D Graph navigation, Euler Operators

1. Introduction

A cell complex may be considered to be made up of closed cells with geometric coordinates at the vertices. A graph connecting the "centres" of the cells is the dual structure to a geometric model. Many researchers consider that only

T. Neutens, P. De Maeyer (eds.), *Developments in 3D Geo-Information Sciences*,
Lecture Notes in Geoinformation and Cartography, DOI 10.1007/978-3-642-04791-6_1,

the geometric graph need be stored, in 2D or 3D, as the dual can readily be reconstructed as required.

This traditional approach has various limitations. In many application disciplines attributes need to be assigned to entities (vertices, edges, faces, volumes) in either the primal or the dual space. (Here we consider the primal space to be the one for which the basic geometric data was acquired.) In some applications both primal and dual entities are needed simultaneously (e.g. watersheds, Maxwell equations). Sometimes the dual graph is critical for computationally-expensive analysis (e.g. network flow through the dual graph, calculation of kinetic Voronoi cell volumes) and it needs to be maintained directly. Sometimes the spatial adjacency of cells needs to be referenced frequently (e.g. flow analysis).

These primal/dual issues may apply in other applications using arbitrary geometry. In particular, CAD systems which permit the modelling of more than one shell (non-manifold geometry) have similar requirements for the data structure. Examples include the Radial-Edge, the Facet-Edge and the Partial-Edge data structures, which require information about loops around vertices, edges and faces. These may be visualized in terms of loops in primal and dual space.

In 2D Guibas and Stolfi (1985) show the advantages of combining the primal and dual graphs in their Quad-Edge (QE) structure. The "Rot" pointer is a combination of the "Sym" pointer of the Half-Edge (HE) structure, which points to the matching Half-Edge, and a "Dual" pointer, which points to the unique dual HE of the current HE. In 3D the Augmented Quad-Edge (AQE) uses the QE for the shell around each vertex in either primal or dual space, but replaces the QE Face pointer by the "Through" pointer, which is equivalent to the "Dual". The AQE allows direct pointer navigation to any entity in either the primal or dual graph.

Model construction is difficult using the AQE because there is no convenient atomic element that can be used for incremental model construction, and construction operators are difficult. It is only implemented for the 3D Voronoi/Delaunay model. In order to facilitate general 3D model construction we developed the Dual Half-Edge (DHE) data structure, which preserves the permanent link between the pair of primal/dual HEs, but allows the reconnection of pairs of HEs during construction. This approach approximately halves the storage with respect to the AQE, and also allows the specification of low-level construction operators that may be combined to produce "cardboard-and-tape" construction methods, 3D Euler operators, or even 2D QEs. These higher level operators may be implemented directly by the model-building software, as in CAD systems. Full navigation of the structure is preserved. Our objective is to use this model for disaster management: the planning of escape routes from complex buildings. Once a primal/dual model is prepared, escape along the edges of the appropriate dual graph (including any relevant navigation attributes, such as speed) may be planned by standard graph traversal methods, and modified in real time if necessary.

We believe that a (relatively) simple 3D primal/dual data structure will be of benefit whenever more than just a geometric model is desired.

2. Related Work

3D data models can be classified into: Constructive Solid Geometry (CSG), boundary-representations (b-rep), regular decomposition, irregular decomposition and non-manifold structures [1]. For our research b-reps and irregular decomposition models are the most relevant. The well known b-reps are: Half-Edge [2], DCEL [3], Winged-Edge [4] and Quad-Edge [5]. Irregular decomposition models (e.g. for constructing a 3D Delaunay tetrahedralization) can be constructed with Half-Faces [6], G-maps [7] and Facet-Edges [8]. The most important for us are the Quad-Edge (QE) and (its extension) the Augmented Quad-Edge (AQE) [1] data structures. These structures are suitable for constructing models and their duals at the same time. We use dual space to connect cells in cell complexes and to navigate between them. Navigation and data structures are the same in both spaces. Both spaces are connected together and we do not need any additional pointers for this connection. Other data structures like Half-Edge or Winged-Edge used widely in CAD systems do not provide for management of the duality.

Quad-Edge The Quad-Edge (QE) was introduced by Guibas and Stolfi [5]. Each QE consists of 3 pointers (Fig 1 a): R, N and V with 4 QEs connected together in a loop by the R pointer to create an edge (Fig. 1 b-c). They are connected in an anticlockwise (CCW) direction. The next pointer N points to the next edge with the same shared vertex or face (Fig. 1b). All edges connected by this pointer form a loop. This is a CCW connection as well. The pointers R and N are directly used in Rot, Sym and Next simple navigation operators. Rot uses R and returns the next quad from a loop of four. Sym calls Rot twice. Next uses N and returns the next edge from the loop (Fig. 1 d). The V pointer is used to point to coordinates of vertices in the structure.

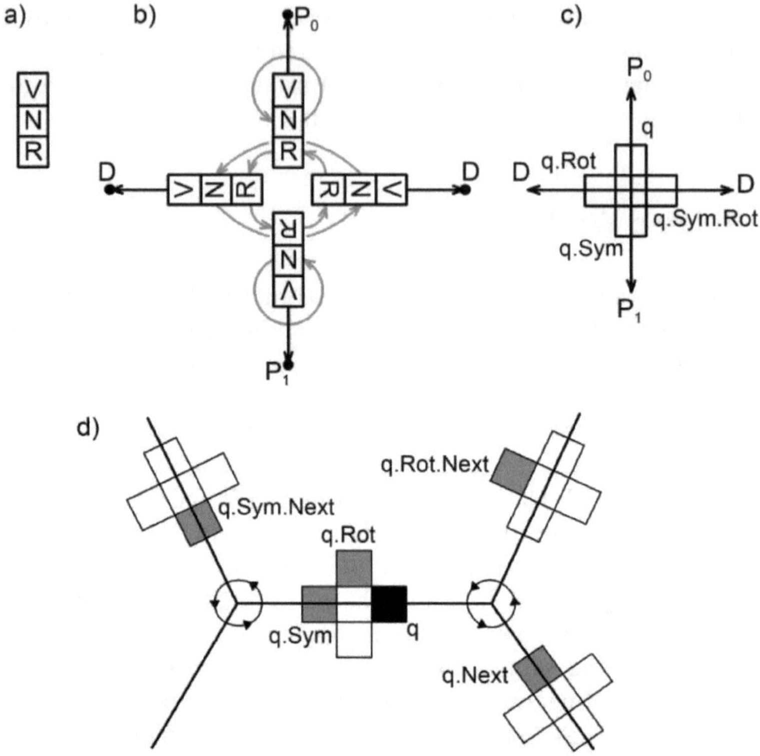

Fig. 1. Quad-Edge structure and navigation; a) single QE; b) four QEs connected using R pointer create an edge P_0P_1; c) simpler representation of the edge P_0P_1; d) relations between edges in a mesh – q (black) represents the original quad, grey quads are an example of navigation.

To construct and modify graphs only two operations are needed – Make-Edge to create a new primal/dual edge pair, and Splice to connect/disconnect it to/from the graph under construction. Tse and Gold showed that Euler Operators could easily be developed on the basis of QEs rather than half or winged-edges [9].

The QE structure was originally used to describe both the primal and dual graphs simultaneously. The particular example was triangulation modelling, showing both the primal Delaunay triangulation and the dual Voronoi tessellation. Either graph may be navigated by simple pointer-following, using Rot and Next pointers, with "Vertex" pointers to primal and dual graph nodes.

Based on the original Quad-Edge, five new developments have led to the current full 3D volumetric Euler Operators:

2.1. Augmented Quad-Edge

In 3D, primal and dual graphs may also be defined: the dual of a face is an edge, the dual of an edge is a face, the dual of a node is a volume and the dual of a volume is a node. Ledoux and Gold defined an extension of the Quad-Edge - the Augmented Quad-Edge (AQE) [1]. In this approach the "Vertex" pointer to a face (usually unused) was assigned to the dual edge of that face. This was called "Through" and allowed navigation between cells via the dual edge.

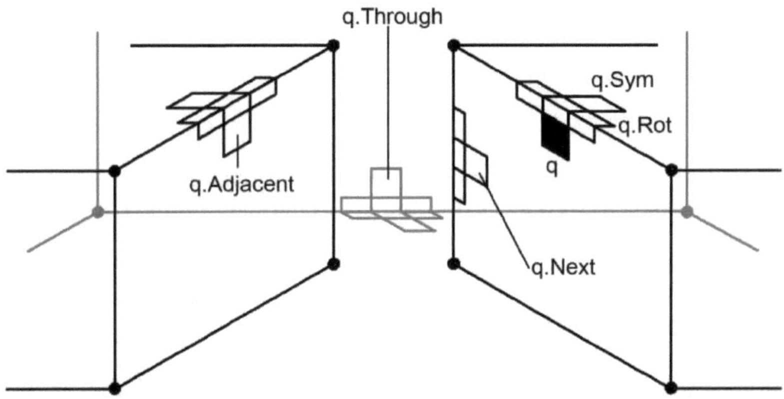

Fig. 2. The AQE structure is suitable for modelling 3D. The set of operators allow for navigation inside one cell as well as between cells.

As seen in Fig. 2, "Through" points to the associated dual edge. Selecting "Sym" then "Through" again returns navigation to the matching edge in the primal space. "Adjacent" is a compound operator returning the matching quad in the adjacent cell. Thus q.Adjacent=q.Through.Next.Through.Sym. The result is the adjacent face from the next cell. The original QE operators are restricted to a single cell; the AQE model allows navigation between cells.

The target application for the AQE was the 3D Voronoi/Delaunay structure. The difficulty with this model was that the construction operators were complex, in particular the "Through" pointer maintenance during VD/DT construction.

2.2. Dual Half-Edge (DHE)

The previous Augmented Quad-Edge was a direct modification of Guibas and Stolfi's [5] Quad-Edge structure. Starting from the AQE, we derived our new Dual Half-Edge structure (Fig. 3) by permitting individual Half-Edges to exist, in order to facilitate construction operations, and enforcing a permanent link between the matching primal and dual Half-Edges, and simplifying the resulting pointer structure. This significantly reduces the space requirements while retaining full navigation of all points, edges, faces and volumes.

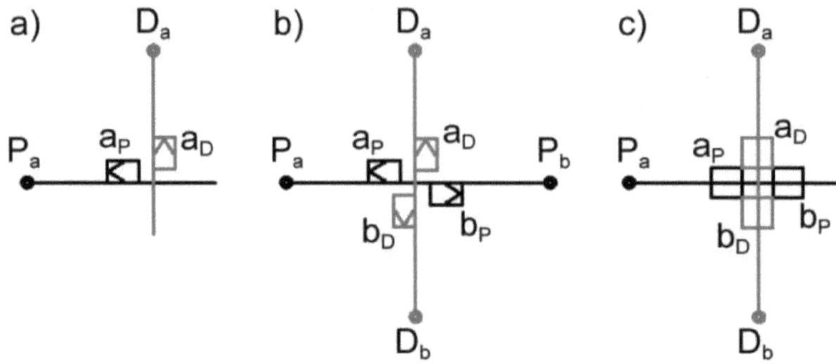

Fig. 3. Dual Half Edge a) singular DHE; the black primal half a_P corresponds to P_a vertex and is permanently connected to the grey dual half a_D that corresponds to the P_{aa} vertex; b) two paired DHEs form an edge; c) the edge represented with QE..

2.3. Atomic elements

In order to create fully navigable structures from the DHE the open "Sym" pointers must be paired to give the required atomic element. A primal or dual edge is formed with one or two distinct vertices. Thus the "Next" pointer has to point to itself or to the second end of the edge. It gives us four possible combinations, as shown in Fig. 4. We use them as a base element for three different construction methods: Quad-Edge (QE), Cardboard&Tape (C&T) and Euler Operators (EO).

The atomic element $P_0P_1D_0$, for example, is the well-known "Make-Edge" element used in the QE structure [5]: two distinct vertices exist in the primal space, but the dual edge connects to itself. Element $P_0D_0D_1$ is the same, except that the two distinct vertices are in the dual space. These are the simplest elements in construction of 2D meshes like triangulation and 3D singular cells. The dual is constructed simultaneously and represents relationships between 2D cells on the 2-manifold. This is equivalent to structures used in simple CAD systems, but cannot be used for a non manifold model.

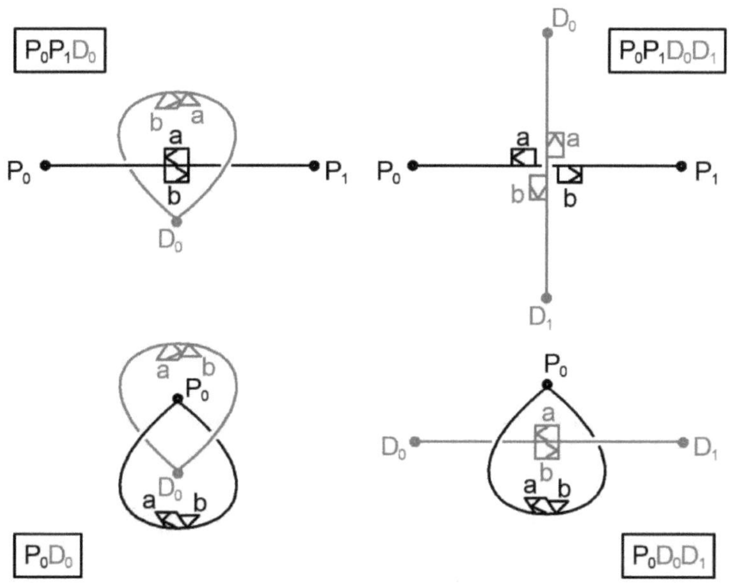

Fig. 4. Four possible combinations of two DHEs. The curved lines are represented in 3D space and form a double-sided face. The face is penetrated by the dual edge.

2.4. Cardboard & Tape models

In [10] we described a spatial model based on directly opposed faces, with no spatial entity between. We called this the "Cardboard and Tape" (C&T) model, as this is intuitively how it is used. To construct a "wall" (the "Cardboard") we start from one vertex and "loop edge" (P_0D_0) and then add new vertices – we split an existing edge. This is illustrated in Fig. 5.

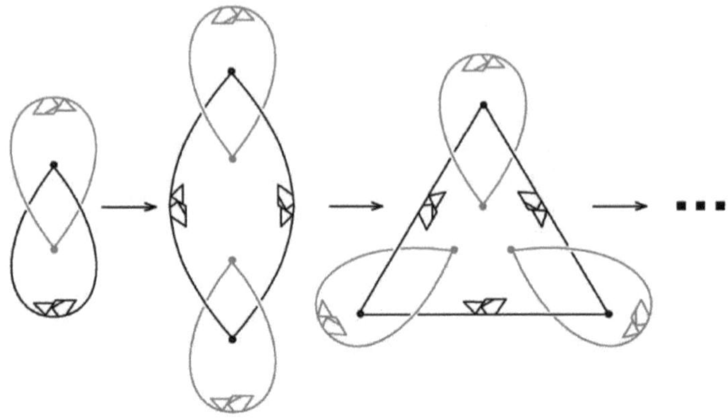

Fig. 5. Construction of a face. New edges are created by adding a new vertex and splitting an existing edge.

These "walls" (or "floors", etc.) are then joined together as desired to give our final building structure, using our "Sew" operator. This simple operator changes only "Sym" pointers in one space. One edge at a time of the double-face is un-snapped, the same is done for the receiving face, and then the half-faces are snapped back together in the new configuration (Fig. 6). This directly mimics the construction of a cardboard model using sticky tape! All operations are defined to be reversible. While not shown in Fig. 6 the dual graph is constructed automatically as the primal graph is built. This approach may be used to construct a cell complex. Visualization of a model is very easy: to get all points, edges or faces of a cell we use a breadth–first traversal of the graph, as we would with a single shell formed by QEs. All cells are connected by the dual structure, which is also a graph.

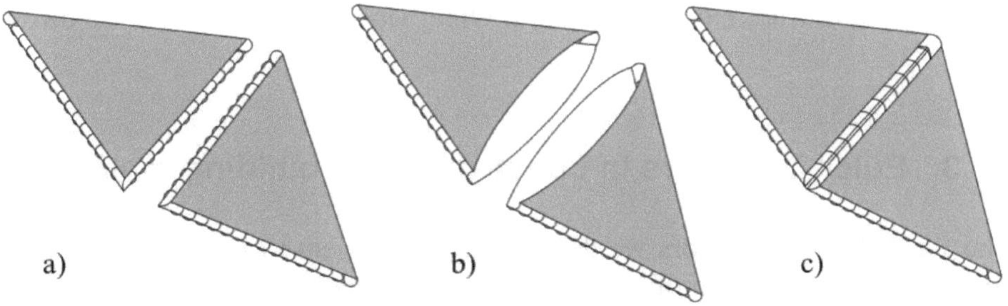

Fig. 6. Sew operator snaps two faces: a) two separate faces; b) unsnapped edges; c) edges snapped into new configuration.

2.5. Euler Operators

The simplest element it is possible to construct in the C&T model is a face. It was not possible to create a single edge with two different vertices (segment line). The primal (geometry) does not cause a problem, but in the dual two not-paired halves of edges do not allow for navigation in the model. The solution to this problem is to keep the external shell together with the internal, original one. This allows us to connect spare pairs of half-edges. It means every edge created using this new method has its counterpart in the external, adjacent shell.

This new idea of keeping the external shell and the last two elements not described yet ($P_0P_1D_0D_1$ and P_0D_0 from Fig. 4) may be used to define an Euler Operator spatial model for full 3D with "Volume" nodes in the dual graph. For complete cell complexes the results are the same as with the C&T model, but it also allows incomplete items, such as edges – which may be needed during the construction process or preserved as final elements. This provides a new data structure for use in CAD system design [11] as an alternative form of non-manifold structure, similar to the radial-edge [12] and facet-edge [8] models.

The main difference between C&T and EO is a possibility of edges construction. Only the EO method allows for that. New edges can be added to the model at any time. In C&T we can add only final faces and connect them with the existing model. Another difference can be found in incomplete models. Two shells (internal and external) are present all the time when the EO method is used. Even a single edge has its external counterpart. It does not make any difference if the created cell is complete or not. In the C&T method the external cell is separated when the cell is closed - for example when the last face of a tetrahedron, cube or any polyhedron is added. The result of these two methods is the same (number of edges, connections between entities, etc.)

only if we construct closed polyhedra. Other (incomplete) models are different.

3. Euler Operators in construction of building models

The set of Euler Operators we developed is sufficient to implement the construction or modification of a solid object. This covers the "spanning set" [13] and is a part of a bigger set of all possible Euler Operators. It may be expanded, and cell complex construction is also possible in the way described by Lee [11].

Construction of a single shell without the dual is a simple process using traditional Euler Operators: the dual may not be important when not working with cell complexes. It should be emphasized here that in our model the dual graph connects individual cells of the primal.

Managing the dual space simultaneously with modification of the primal requires more complex operators – although once defined they may be used as simply as the traditional ones. In return we get automatic and local changes in the dual when the geometry of the primal is changed. No additional operations are needed. The dual is present all the time and navigation in the model is valid at every step of the construction process.

The construction process for a single triangle is shown in Fig. 7. In order to manipulate individual edges as in traditional Euler methods (as opposed to the complete faces described above for the Cardboard and Tape model) we have found that we need to work with double shells (interior and exterior) throughout. Two $P_0P_1D_0D_1$ elements of Fig. 4 are spliced to give the basic element of Fig. 7. These are then joined to give the triangle of Fig. 7: there is an interior and exterior triangle in 3D space, and the equivalent in the dual. (A similar process may be developed using P_0D_0.)

The dual to an internal or external triangle is respectively an internal or external node. In fig. 4 dual nodes were 'split' to show that dual cells (grey edges) are not connected directly (dual cells are connected by primal edges). In fact there are only two dual nodes present during the triangle construction process – one corresponds to the internal triangle and another one corresponds to the external triangle. These two nodes are present in the model even if the triangle is not constructed completely (is not closed).

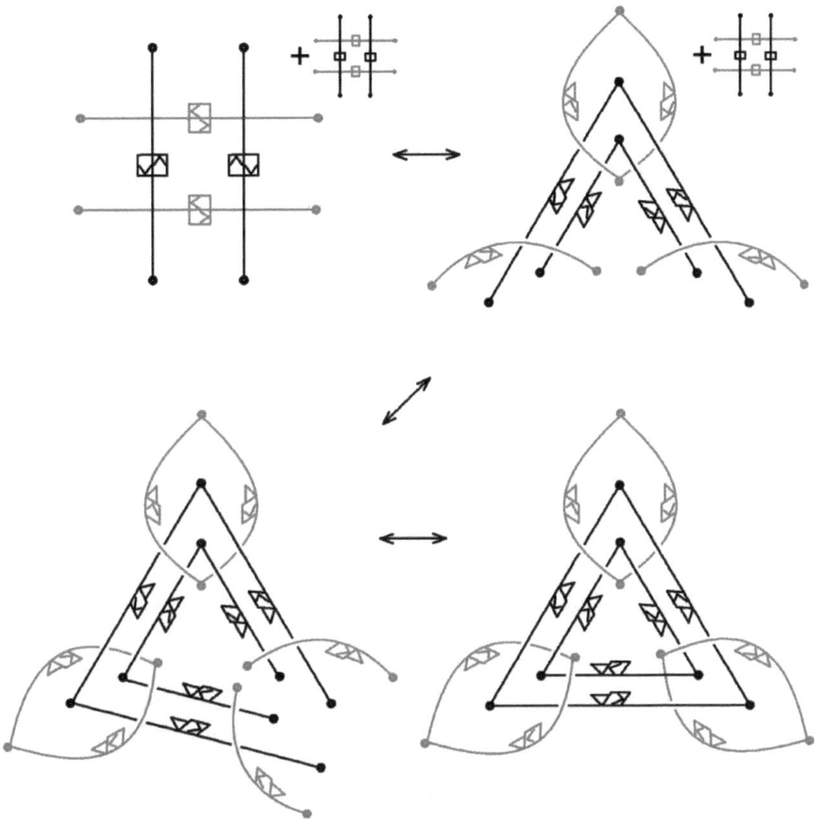

Fig. 7. Construction of a triangle in 3D in a model based on the Euler operators. Internal and external primal shells (black edges) are connected by the dual (gray edges).

Using standard Euler operators we can easily build shells of any shape (except shells with holes, which are not yet included). Fig. 8 shows the sequence used to create a cube from individual edge elements. (This is one of many possible sequences.) For clarity the dual is not shown, but it is present at each step, as with the external shell. The dual connects the internal and external cells together into one cell complex. Faces are defined automatically upon closure of the edges, and volumes are determined whenever a closed set of faces is completed.

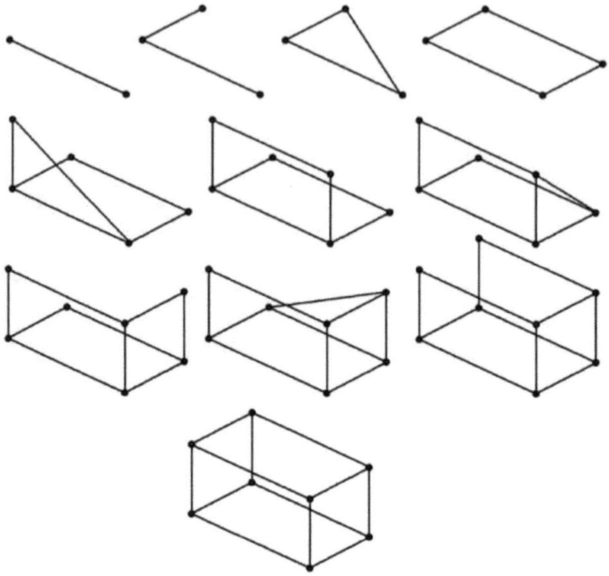

Fig. 8. One of many possible ways of cube construction using Euler Operators.

The cell complex consisting of two cells (internal and external) is not very useful. For practical models, we need to be able to construct more complex structures. We extend the set of standard Euler operators to include operators modifying not only a single shell but also a cell complex. Using them we can connect/disconnect or merge/split cells of the complex. Fig. 9 presents this idea. In fact we perform a sequence of very simple, atomic operations, the same as we use with standard Euler operators. Lee [11] shows a similar example using standard Euler Operators (with no dual) for joining or splitting cells.

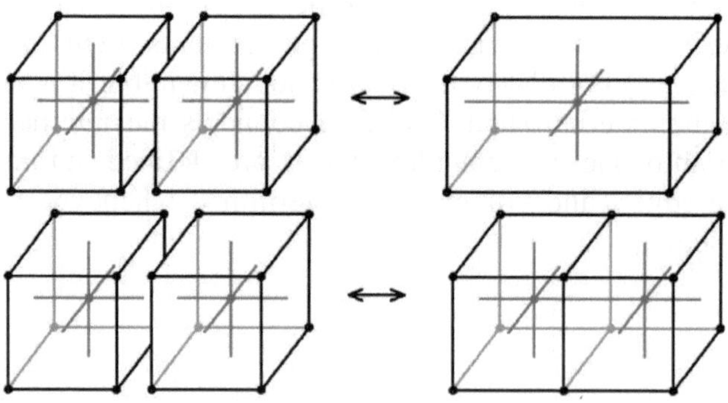

Fig. 9. Merge/Split and Connect/Disconnect operations on elements of a cell complex.

It should not be difficult to imagine the construction of more complex structures. We can easily connect shells by adjacent faces, but the real world is not so simple. Not every wall in a building is shared by exactly two rooms, e.g.: one long corridor may have many neighbouring rooms. Neighbouring rooms can have walls of different shape or size. To use our idea for more practical applications we had to develop a boundary intersection module. During the construction process we check the locations of two adjacent shells, add new edges if necessary, and then connect them (Fig. 10). We do not test to see if shells overlap. This can be a subject for future work.

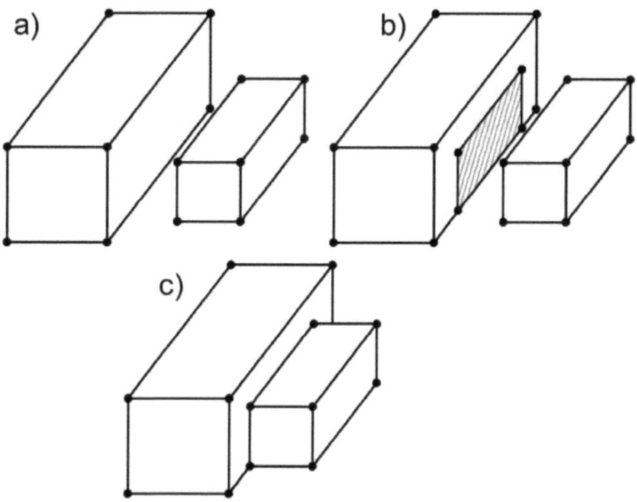

Fig. 10. Boundary intersection module. a) two adjacent shells; b) new edges added to a bigger face create a new adjacent face; c) two shells connected.

4. Emergency management systems and navigation of building models

Geometry, topology and information are three base elements in a CAD model [14]. Geometry and topology are included in the DHE data structure, and can be changed using our operators. Information is easy to implement using attributes assigned to entities (i.e. vertices and half-edges). Because our data structure is a pointer based structure, it is easy to extend it to manage additional attributes. Because of the duality of our models, attributes assigned to a node correspond with a dual volume; attributes assigned to an edge may be treated as if they are connected with a dual face. These attributes can be used in various ways, depending on the application. For example we can use them as weights for graph traversal, or to find the shortest path between two nodes.

We can solve the problem of finding escape routes from buildings using our structures and appropriate graph algorithms. If the primal represents the geometry of rooms in a building, the dual will be the graph of the connections between the rooms. Nodes in the dual represent rooms, and edges connect adjacent rooms. Now we only use the dual to make all calculations. We mark some of the dual nodes of a model with an attribute 'room' (and some will also be considered 'exits'). All connections between separate rooms have an attribute 'distance' with a weight value. The higher the weight, the harder it is to pass the connection. All connections that are impossible to pass (e.g. a ceiling, a thick wall) have an infinite weight. We start our searching from the node 'room' (dual to the node is a cell in the primal that describes the geometry of the room) and we try to find the route to the closest 'exit'. Then we calculate routes for the next room and so on. We use the Dijkstra algorithm.

We reconstructed a simple model of a building presented by Lee in [5] (Fig. 11 a) with methods described in the previous section (Euler operators). The geometrical model of the building (Fig. 11 b) is a set of shells connected by the dual structure (Fig. 11 c). The shells are marked with S1-S12 symbols, and each represents a single room, corridor or staircase. Not all connections between shells can be passed. For example there is no passage through ceilings or walls. We set the weight attributes for such connections. The final graph of accessible connections is presented in Fig. 11 d) and can be used in calculations of escape routes.

We have shown this model before in [10]. There we used a slightly different data structure with Cardboard & Tape operators. The final model is the same, but by using Euler operators and the intersection module our methods are more universal. They could be used in CAD systems to build multi-shelled objects with included topology.

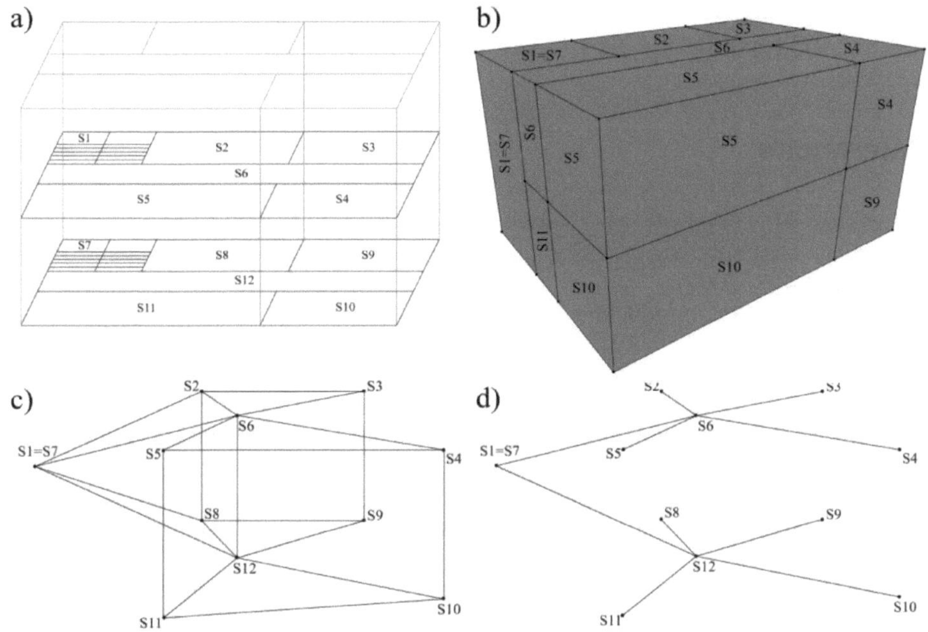

Fig. 11. Example structure representing a 3D model of a building interior. S1, S7 – staircase, S2-S5, S8-S11 rooms, S6, S12 – corridor: a) spatial schema – original version [5], b) volumetric model of rooms, c) complete graph of connections between rooms, d) graph of accessible connections between rooms.

5. Conclusions

The DHE data structure and methods described for the construction of 3D models may be useful where both geometry and topology are important for cell complexes. Models are represented by connected dual graphs. Emergency management and modelling of building interiors in 3D are possible applications for our models. The geometry of a building is kept in the primal graph and the topology (connections between rooms) – in the dual. Using the dual and the Dijkstra graph traversal algorithm, searching for escape routes from buildings can be implemented to plan rescue operations. There are other examples, where the duality of graphs has significant meaning: for example to solve the discrete Maxwell equations for electromagnetic radiation [15]. Use of the dual graph in CAD systems appears to be new (Kunwoo Lee, November 2008, personal communication).

Acknowledgments This research is supported by the Ordnance Survey and EPSRC funding of a New CASE award.

References

1 Ledoux, H., Gold, C. M.: Simultaneous storage of primal and dual three-dimensional subdivisions. Computers, Environment and Urban Systems 31 (4), pp. 393-408 (2007)

2 Mantyla, M.: An introduction to solid modelling. Computer Science Press, New York, USA (1998)

3 Muller, D. E., Preparata, F. P.: Finding the intersection of two convex polyhedra. Theoretical Computer Science, vol. 7, pp. 217-236 (1978)

4 Baumgart, B. G.: A polyhedron representation for computer vision. In: National Computer Conference, AFIPS (1975)

5 Guibas, L. J., Stolfi, J.: Primitives for the manipulation of general subdivisions and the computation of Voronoi diagrams, ACM Transactions on Graphics, vol. 4, pp. 74-123 (1985)

6 Lopes, H., Tavares, G.: Structural operators for modelling 3-manifolds. In: Preceedings 4th ACM Symposium on Solid Modeling and Applications, Atlanta, Georgia, USA, pp. 10-18 (1997)

7 Lienhardt, P.: N-dimensional generalized combinatorial maps and cellular quasi-manifolds. International Journal of Computational Geometry and Applications, vol. 4 (3), pp. 275-324 (1994)

8 Dobkin, D. P., Laszlo, M. J.: Primitives for the manipulation of three-dimensional subdivisions. Algorithmica, vol. 4, pp. 3-32 (1989)

9 Tse, R. O. C., Gold, C. M.: TIN Meets CAD - Extending the TIN Concept in GIS. Future Generation Computer systems (Geocomputation), vol. 20(7), pp. 1171-1184 (2004)

10 Boguslawski, P., Gold, C.: Construction Operators for Modelling 3D Objects and Dual Navigation Structures, in: Lectures notes in geoinformation and cartography: 3d Geo-Information Sciences, Part II, S. Zlatanova and J. Lee (Eds.), Springer, p. 47-59 (2009)

11 Lee, K.: Principles of CAD/CAM/CAE system, Addison-Wesley/Longman, Reading (1999)

12 Weiler, K.: The Radial Edge Structure: A Topological Representation for Non-manifold Geometric Boundary Modeling, in Geometric Modeling for CAD Applications, Elsevier Science (1988)

13 Braid, I. C., Hillyard, R. C., Stroud, I. A.: Stepwise construction of polyhedra in geometric modelling, in: Mathematical Methods in Computer Graphics and Design, ed. K W. Brodlie, Academia Press (1980)

14 Stroud, I.: Boundary Representation Modelling Techniques, Springer (2006)

15 Sazanov, I., Hassan, O., Morgan, K., Weatherill, N. P.:Generating the Voronoi – Delaunay Dual Diagram for Co-Volume Integration Schemes. In: The 4th International Symposium on Voronoi Diagrams in Science and Engineering 2007 (ISVD 2007), pp. 199-204 (2007)

True-3D Visualization of Glacier Retreat in the Dachstein Massif, Austria: Cross-Media Hard- and Softcopy Displays

Katharina Bruhm, Manfred Buchroithner and Bernd Hetze

Dresden University of Technology, Germany
*Corresponding author: manfred.buchroithner@tu-dresden.de

Abstract. Glacier recession is a global phenomenon subject to climate change. This also applies to the Dachstein Massif in die Eastern Alps of Austria. Based on historical and recent maps, and moraine mapping the glacier states from the years 1850, 1915 and 2002 were used as input for for photorealistic reconstructions and visualizations of the respective glacier states. A detailed digital terrain model and aerial photographs (2003 – 2006) were provided by the Government of Styria and Joanneum Research Graz. By means of the software packages ERDAS *Imagine 9.1*, ESRI *ArcGIS 9.2*, 3D Nature *Visual Nature Studio 3* (*VNS*), Digi-Art *3DZ Extreme* and Avaron *Tucan 7.2* the glacier conditions during the "Little Ice Age" (+/- 1850) and the following two dates were reconstructed. Subsequently, several derivates of these data sets were generated.

First, three individual overflight simulations were computed, permitting to obtain a realistic impression of the Dachstein Massif and its glaciers in 1850, 1915 and 2002. As a second embodiment product, a fast-motion dynamic visualization of the glacier recession was generated which illustrates their decrease in thickness. Third, combining both the flip effect and the true-3D effect achievable by lenticular foils, were applied to produce a multitemporal autostereoscopic hardcopy display. Fourth, the overflight simulation data sets were used to generate stereo-films which could then be displayed on back-projection facilities using either passive polarization glasses or active shutter glasses.

T. Neutens, P. De Maeyer (eds.), *Developments in 3D Geo-Information Sciences*,
Lecture Notes in Geoinformation and Cartography, DOI 10.1007/978-3-642-04791-6_2,

Introduction

Glacier recession is a global phenomenon subject to climate change. This also applies to the Dachstein Massif in die Eastern Alps of Austria. The Massif is situated in the Federal States Styria, Salzburg and Upper Austria and covers an area of nearly 900 square kilometers (cf. Figure 1). The High Dachstein is with 2.995 meters the highest point in this region. On this plateau range lay nine glaciers. As part of the UNESCO World Nature and Culture Heritage *Hallstatt – Dachstein / sSalzkammergut* and also as an important factor for the economy the retreat of the glaciers is monitored with concern. The project described in this paper is meant to show the changes of the glacier coverage over the last 150 years, a phenomenon which is gaining increasing important in times of global warming.

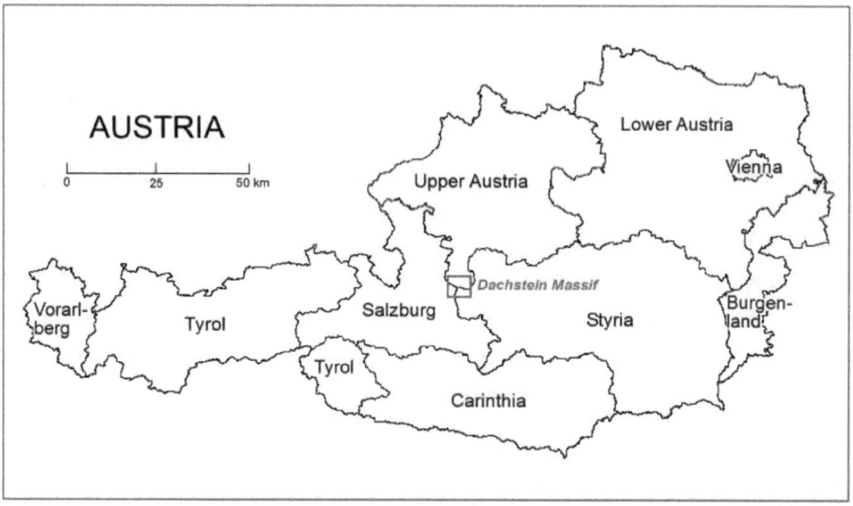

Fig. 1. Map of Austria in 1 : 2.800.000 showing location of the Dachstein Massif.

Based on historical and recent maps, and moraine mappings based on aerial photography the glacier states of 1850, 1915 and 2002 were used as input for photorealistic reconstructions and visualizations of the respective conditions. A detailed digital terrain model and aerial photographs (2003 – 2006) were provided by the Government of Styria and Joanneum Research

Graz. By means of the software packages ERDAS *Imagine 9.1*, ESRI *ArcGIS 9.2*, 3D Nature *Visual Nature Studio 3* (*VNS*), Digi-Art *3DZ Extreme V7,* Awaron *Tucan 7.2* the glacier conditions during the "Little Ice Age" (+/- 1850) and the following two dates were reconstructed. Subsequently, several products of these data sets were generated, thus taking up the idea of cross-media visualization, however in stereo-vision.

Overflights with Visual Nature Studio

As first product, three individual overflight simulations were computed, permitting to obtain a realistic impression of the Dachstein Massif and its glaciers in 1850, 1915 and 2002. The conditions in the Dachstein Massif at the end of the "Little Ice Age" are only recorded in drawings and panoromic depictions by the Austrian geoscientist Prof. Friedrich Simony who explored this area between 1840 and approx. 1890. An accurate map at a scale of 1 : 10.000 was made by Arthur von Hübl in 1901 showing the Hallstätter Glacier, the largest of the Massif, in the year 1899. An analysis of moraines carried out by Michael Krobath and Gerhard Lieb from the University of Graz, Austria (Krobath & Lieb, 2001) was used for the mapping of all major Dachstein glaciers. The "virtual overflights" are supposed to give an impressive picture of the last maximum of the glaciers at around 1850 and their subsequent retreat.

At the beginning of the 20th century the Austrian and German Alpine Club (Alpenverein) began to explore the Dachstein Massif. Their "Alpenverein Map" published in 1915 contains a highly detailed depiction of the glaciological conditions. It reveals an explicit glacier retreat of approx. 30 % of the glacier area within the past 65 years (cf. Table 1). The glacier tongues receded above 2000 meter asl, implying a reduction of over 25 percent of their length. The most significant change was the separation of the Little Gosau Glacier from the Northern Torstein Glacier. Over 50 percent of there area were lost (cf. Figure 2).

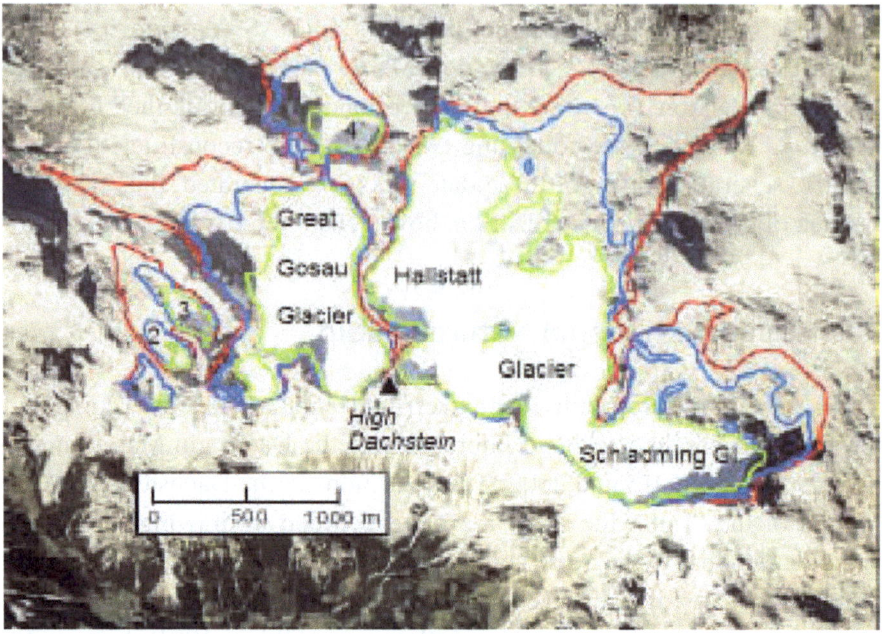

Fig. 2. Orthophoto mosaic of the Dachstein Massif showing selected glacier states. Red: 1850, blue: 1915, green: 2002. 1: Southern Torstein Glacier, 2: Northern Torstein Glacier, 3: Little Gosau Glacier, 4: Schneeloch Glacier.

Since 1946 the Alpine Club undertook continuous annual measurements of the glacier tongues and frequent surveys of their surfaces. The last accurate map was published in 2005 displaying the glacier state of 2002. Since 1850 the appearance of the area has been changing significantly: About 50 % of the glacier surfaces got lost. The snouts of the glaciers are now ending at an altitude higher than 2150 meter, and the recession process is still going on (cf. Table 1, Krobath & Lieb, 2001).

Table 1. Glacier surfaces of 1850, 1915 and 2002 in square meters.

Glaciers	1850	1915	2002
Hallstätter / Schladminger Gl.	7568460	6114510	3983797
Great Gosau Glacier	2559852	2007971	1312897
Little Gosau Glacier	472440	160530	95882
Northern Torstein Glacier		88288	25750
Southern Torstein Glacier	65107	82581	4668
Schneeloch Glacier	678195	458242	169478
Total [sqm]	11.344.054	8.912.122	5.592.472
Area loss since 1850 [%]		23,8	49,3

A quick construction of the three glacier states was realized with the 3D Nature *Visual Nature Studio*. This software package offers the opportunity to import various GIS or geospatial data like CAD DXF, *Arc* ASCII-files, several images formats (Tiff, JPEG, …) and vector shape-files in the selected geometric reference system. The materialization is quite easy: First the terrain data were loaded as an ASCII-file and then texturized with the air photo mosaic. Based on the Alpine Club Maps of 1915 and 2002 and the publication of Krobath & Lieb (2001) displaying the glacier state of 1850 the necessary vector information was digitized. The Alpine Club Maps represent the most accurate glacier information for abovementioned years.

Subsequently the imported vectors were used for the glacier modeling. On one hand they were utilized to adapt the digital terrain model within the vector polygons. Images containing the z values of the glacier (in the form of grey values) increase or decrease the ground inside the shape files with the help of the special tool *area terraffector*. In these images the differences between the glacier models and the digital terrain model are shown. With the help of the Erdas Imaging Model Maker these pictures of all glaciers were separated into negative and positive areas above and below zero meters of change. Thus, two images per glacier were calculated. Furthermore, the vector polygons are texturized as snow, by means of the *snow effect* tool.

With only few processing steps three rather photorealistic models of the Dachstein glaciers of different points in time were created (cf. Figures 3 and 4).

A close look at the glaciers, however, reveals that the typical marginal bulges of the glaciers are missing. In VNS glacier margins cannot be represented in their typical photorealistic way. This would imply the use of extra software. In order to further model the 3D models an export into a more general 3D data format like PRJ is needed. These models can then be displayed using 3D Studio or similar programmes, however, a decrease in visual quality and accuracy has to be accepted. This can be explained the limited possibilities provided by the export settings of VNS. Due to time reasons the glacier margins have thus not been altered. However, for the animation of the glacier downwasting, i. e. the reduction in thickness, (see below) a particular modelling the glacier margins has been performed.

Fig. 3. Dachstein glaciers in 1850 – view from the North.

Fig. 4. Dachstein glaciers in 2002 – view from the North.

Animation of ice thickness reduction

As a second embodiment product, a fast-motion dynamic visualization of the glacier recession was generated which – due to some slight vertical exaggeration – also effectively illustrates their decrease in thickness, the so-called downwasting. In 1968 the Federal Office of Meteorology and Geodynamics in Vienna performed seismic measurements at the Dachstein glaciers in order to determine the average ice thicknesses (Brückl et al., 1971). These measurements, repeated in 2000 with the help of GPS, form the basis for the representation of the inferred thickness around the year 1850. For the years 1915 and 2002, for every glacier digital terrain models were manually generated. These models then represented the bases for the calculation of the downwasting rates using the 2004 digital terrain model generated by the Joanneum Research Graz. This model, provided by Government of Styria, has a resolution of 10 meters and was derived from the 1 : 50.000 topographic maps. The results of the calculation represent the differences between the models of 1915 and 2002 respectively and the recent terrain model of Styria, i. e. the relative changes of the glacier surface morphology in z direction. Hence, the aforementioned animation, realized by means of *Tucan*, also shows a significant reduction in glacier thickness over the last 150 years.

The Tucan software was used for the representation of the 3D model computed in real time. By means of this software the "unintelligent" geometric input data were converted into complex scenes originates (cf. Figure 5).

Both free navigation in a scene and the movement along a pre-defined path are possible. Besides a free viewer, for the displaying of virtual scenes also a professional variant is available. Active and passive stereo-vision are supported.

The software was designed in such a way that it is simply applicable by users without Virtual Reality expertise and programming knowledge. Tucan offers many functions for the production even of complicated animations. Based on an integrated Interface Designer it is possible to steer moving geometries as well as dynamic and particular system settings e.g. by buttons (www.awaron.com/en/products/tucan). This functionality was used for the representation of the glacier changes.

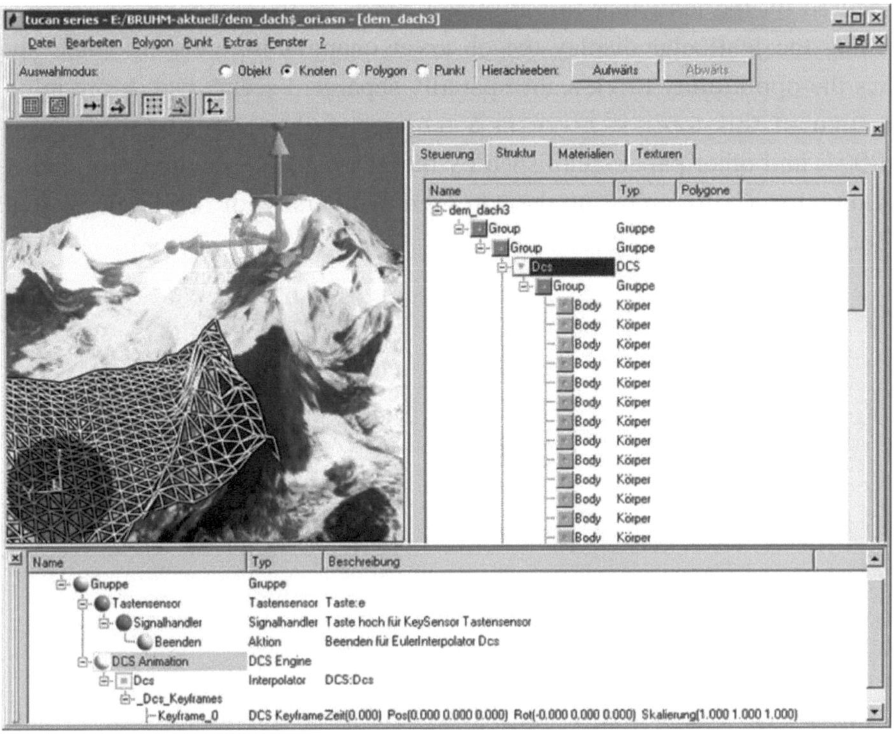

Fig. 5. Screenshot showing the Tucan software work environment.

Lenticular display

Third, both the flip effect and the true-3D effect achievable by lenticular foils, were combined to produce a multitemporal autostereoscopic hardcopy display. Lenticular foils offer a nice alternative to demonstrate dynamic processes in the form portable hardcopies viewable with unaided eyes.

The basic principle of the lenticular method is the artificial stereoscopic viewing created with the help of semi-cylindrical parallel micro lenses running on the top side of a transparent synthetic foil. Under these lenses interlaced images are arranged which means under every lens lies one pixel column of all used stereo-mates (cf. Figure 6). For stereoscopic viewing the left and the right eye need images projected in different directions. The foil permits the separation of these stereo-mates. Incoming optical rays are focused by the lenticular foil at several strips of the image. By turning or tilting the lenticular display the observer changes his viewing angle and has the opportunity to view the spatially separated scene data. More information of this topic is given in a monograph by Thomas Gründemann (2004) and other papers published by the 3D visualization group of the Institute for Cartography at the Dresden University of Technology (e.g. Buchroithner et al. 2005a, 2005b).

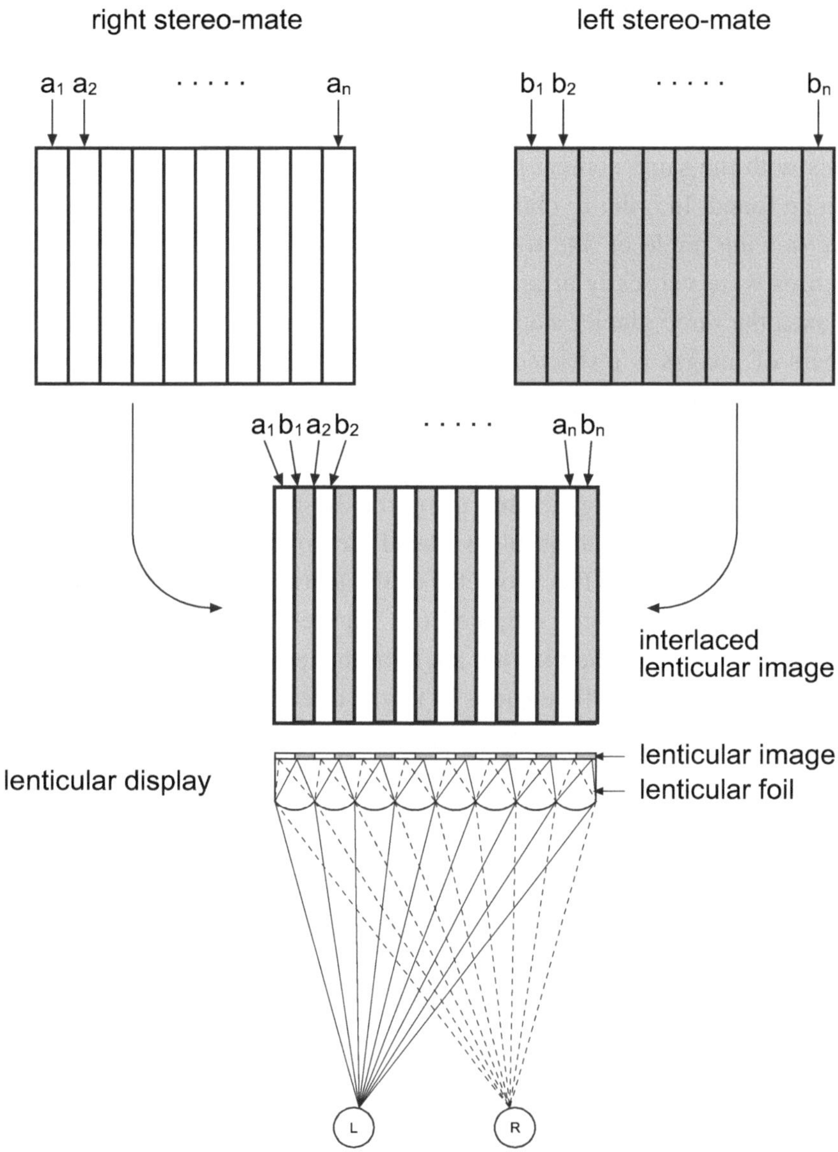

Fig. 6. Principle of lenticular foil method showing the interlacing of stereo-images.

Display possibilities of the lenticular method are 2D effects like flipping, morphing or animation and the stereoscopic 3D effect. The combination of 2D and 3D effects is also possible. Such a combination of flip and true-3D effect is rather complex. The flip effect is the easiest type of the effects.

Two or more stereo-images are used. By tilting the display the viewer can only see one of the scenes. Generally the micro lenses are, in contrast to the true-3D stereo-vision effect, horizontally arranged. Here the lenticular lenses run in vertical direction. The effect is generated by using different images with the same content but different view. Therefore a compromise has to be found. In order to obtain a good result foils with 62 lines per inch and a viewing angle of 44° were used. For a sufficient depth impression the lenses were vertically arranged. By tilting the hardcopy along the vertical axes the three glacier states of 1850, 1915 and 2002 are visible. The numbers of images is restricted to three stereo-scenes. To avoid a kind of fusion of the different images, called cross-talking or ghosting, two neutral ("information-free") images "in between" the three scenes are required.

The stereo-images were generated using the 3D-modeling software *Visual Nature Studio* by simulating an overhead stereo-camera with parameters especially calculated for this project. For the lenticular display two images of every glacier date were computed (cf. Figures 7 and 8). Additional to the numbers of images the viewing angle of the lenticular foil, the horizontal field of view and the dimensions in x, y, z direction of the digital terrain model were needed. The calculated parameters were e.g. the focal length, the distance of the camera above ground and the image base. After the rendering these images are taken and served as input for further processing by the lenticular software *3DZ Extreme V7*. 3DZ is an advanced and easy-to-use lenticular and barrier screen printing software (www.digi-art.de). It was used to interlace the images. The final product has a size of 25 x 16 cm.

Fig. 7. Overhead view of 1850 glacier simulation, left stereo-mate.

Fig. 8. Overhead view of 1850 glacier simulation, right stereo-mate.

Stereo-overflights for back-projection facilities

The overflight simulation data sets were used to generate stereo-overflights by *Tucan* which were then displayed on back-projection facilities. These walkthroughs, rendered in real-time, are shown trough a translucent screen to the spectators who observe the displayed scenes from the other side of the screen. Using either passive polarization glasses for stationary back-projection facilities or active shutter glasses for portable ones, this technique is unfortunately not "auto"-stereoscopic. At the Dresden University of Technology both types of back-projection facilities are used for teaching purposes.

This technology achieves a so far unequaled quality in stereovision. This is certainly subject to the high resolution of the used projectors which is currently in the range of 1400 x 1050 pixels for portable and 1920 x 1200 pixels for stationary high-end devices. For the control of the graphics periphery an NVIDIA Quadro graphic card is used.

The Digital Terrain Model of the present test-site comprehends 1.600.000 polygons which, for the reason of real-time visualization, had to be converted into an elevation grid. Even the total number of polygons covering glaciated terrain amounts to more than one million.

Conclusion

All the above mentioned products are based on the same data sets. In the sense of cross-media visualization they are planned to be presented to both the scientific and the public communities at various occasions. Initial internal tests with students already revealed the high potential of these individual products for a better understanding of the still ongoing glacier retreat in the Alps.

To the authors' knowledge all the above described stereo-visualization methods have never before been applied to real and simulated – in our case historical – geodata. This premiere is meant to demonstrate different approaches of stereo-visualization for environmental purposes. Furthermore, the applied display techniques are supposed to give an impression of the great variety of 3D viewing hard- and softcopy methods.

Due to the different media forms used, identical data bases can be presented to different user groups. Overflight simulations and lenticular foil displays proved to be particularly well suited for tourism purposes. Animations might help to improve the understanding of complex geo-processes.

Despite their restricted possibilities for the presentation of dynamic processes, lenticular foil hardcopy displays have significant advantages for outdoor use. They require neither viewing means nor electric power. The applied softcopy technology needs, albeit, stereo-glasses but it allows the visualization of photorealistic scenes with a so far unequaled quality, also in animated form.

A comparative assessment of the two used 3D software packages Visual Nature Studio and Tucan is given in Table 2.

Table 2. Pros and cons of the applied 3D software packages. Evaluation range from - - to + +.

Software package	VNS	Tucan
Easiness of use	+ +	+
Generation of complex animations	-	+
Integration of GIS data	+ +	-
Import/Export	o	o

References

Brückl E, Gangl G, Steinhauser P (1971) Die Ergebnisse der seismischen Gletschermessungen am Dachstein im Jahre 1968. Arbeiten aus Zentralanstalt für Meteorologie und Geodynamik, Vol. 9, Wien

Buchroithner MF, Gründemann T, Randolph KL, Habermann K (2005a) Three in one: Multiscale Hardcopy Depiction of the Mars Surface in Trur-3D. Photogrammetric Engineering & Remote Sensing Okt. 2005 : 1105 - 1108

Buchroithner MF, Habermann K, Gründemann T (2005b) Modeling of Three-Dimensional Geodata Sets for True-3D Lenticular Foil Displays. Photogrammetrie – Fernerkundung – Geoinformation 2005/1 : 47 - 56

Gründemann T (2004) Grundlagenuntersuchungen zur kartographischen Echt-3D-Visualisierung mittels des Lentikularverfahrens. Pre-Thesis, Dresden University of Technology

Habermann K (2004) Conception for True-3D Maps. Presentation at the XIII. Cartographic School 2004 of University Breslau: Co zwie si□ koncepcj□ mapy? - What is called the Map Conception?. pp. 76 – 88.

Krobath M, Lieb G (2001) Die Dachsteingletscher im 20. Jahrhundert. K. Brunner (Editor), Das Karls-Eisfeld. Research Project at Hallstätter Glacier. Wissenschaftliche Alpenvereinshefte 38: 75 - 101

Mandl F (2006) Der Hallstätter Gletscher. Alte und neue Forschung. ALPEN – Festschrift: 25 Jahre ANISA : 202 - 215

Simony F(1885) Ueber die Schwankungen in der räumlichen Ausdehnung der Gletscher des Dachsteingebiges während der Periode 1840 – 1884. Mittheilungen der kais. königl. Geographischen Gesellschaft in Wien. Vol. 28 : 113 - 135

Simony F (1889). „Das Dachsteingebiet". Hölzel, Wien-Olmütz

3D Nature (2009). www.3dnature.com

Awaron Tucan (2009) www.awaron.com/en/products/tucan

Digi-Art (2009) www.digi-art.de

Planai (2009) www.dachstein.planai.at

Towards Advanced and Interactive Web Perspective View Services

Benjamin Hagedorn, Dieter Hildebrandt and Jürgen Döllner

Hasso-Plattner-Institute at the University of Potsdam,
Prof.-Dr.-Helmert-Str. 2-3, 14482 Potsdam, Germany,
{benjamin.hagedorn|dieter.hildebrandt|doellner}@hpi.uni-potsdam.de

Abstract. The Web Perspective View Service (WPVS) generates 2D images of perspective views of 3D geovirtual environments (e.g., virtual 3D city models) and represents one fundamental class of portrayal services. As key advantage, this image-based approach can be deployed across arbitrary networks due to server-side 3D rendering and 3D model management. However, restricted visualization and interaction capabilities of WPVS-based applications represent its main weaknesses. To overcome these limitations, we present the concept and an implementation of the *WPVS++*, a WPVS extension, which provides A) additional thematic information layers for generated images and B) additional service operations for requesting spatial and thematic information. Based on these functional extensions, *WPVS++* clients can implement various 3D visualization and interaction features without changing the underlying working principle, which leads to an increased degree of interactivity and is demonstrated by prototypic web-based client applications.

1. Introduction

Within spatial data infrastructures, geovisualization plays an important role as it allows humans to understand complex spatial settings and to integrate heterogeneous geodata from distributed sources on the visualization level. For this purpose, the Open Geospatial Consortium (OGC) as a standardization organization proposes several portrayal services for 2D and 3D geodata. So far, there is only one widely used "workhorse", the Web Map Service (WMS) [4], for generating 2D maps. Standards for visualizing 3D geovirtual environments (3DGeoVEs) such as virtual 3D city models and

T. Neutens, P. De Maeyer (eds.), *Developments in 3D Geo-Information Sciences*,
Lecture Notes in Geoinformation and Cartography, DOI 10.1007/978-3-642-04791-6_3,

landscape models have not been elaborated to a similar degree. For portraying 3D geodata, two approaches have been presented: the Web 3D Service (W3DS) and the Web Perspective View Service (WPVS).[1]

While the W3DS provides scene graphs that have to be rendered by the service consumer, the WPVS supports the generation of images that show 2D perspective views of 3DGeoVEs encoded in standard image formats. The key advantages of this image-based approach include low hardware and software requirements on the service consumer side due to service-side 3D model management and 3D rendering, and moderate bandwidth requirements; only images need to be transferred regardless of the model and rendering complexity. The weaknesses of the current WPVS approach include the limited degree of interactivity of client applications. For example, it is not possible to navigate to a position identified in the image or to retrieve information about visible geographic objects (e.g., buildings).

To overcome these limitations, we propose the *WPVS++*, an extension of the OGC WPVS that is capable of augmenting each generated color image by multiple thematic information layers encoded as images. These additional image layers provide for instance depth information and object identity information for each pixel of the color image. Additionally, we propose operations for retrieving thematic information about presented objects at a specific image position, simple measurement functionality, and enhanced navigation support. This allows *WPVS++* clients to efficiently access information about visually encoded objects in images, to use that information for advanced and assistant 3D navigation techniques, and thereby to increase the degree of interactivity, which is demonstrated by prototype implementations of two web-based clients.

The remainder of this paper is organized as follows. Section 2 describes and analysis approaches for service-based 3D portrayal and reviews related work. Section 3 presents the extended information model of the *WPVS++*; the extended interaction model is proposed in Section 4. Section 5 describes our prototype implementation. Section 6 discusses results, and Section 7 concludes the paper.

[1] The W3DS has discussion status. The WPVS is an OGC-internal draft specification and represents the successor of the OGC Web Terrain Server discussion paper [15].

2. Background and Related Work

2.1. Service-Based 3D Portrayal

The OGC portrayal model [12] allows for three principle approaches for distributing the tasks of the general rendering pipeline between portrayal services and consuming applications [1]. These segmentations differ in the type of exchanged data: either filtered geodata, or computer graphical representations, or finally rendered images are provided. This leads to different rendering capabilities and hardware and software requirements of the portrayal services and the corresponding client applications.

The *OGC Web 3D Service* (W3DS) [13] generates 3D computer graphical representations, which are typically arranged as *scene graphs* and are transmitted to the service consumer using 3D graphics formats (e.g., VRML, X3D, COLLADA). This graphics data can include appearance information, e.g., surface colors, but is not capable of carrying thematic information. These representations have to be processed and rendered at the client side. Using the W3DS together with a powerful rendering client allows for real-time navigation and interaction in the 3D scene. However, for the provision of complex scenes by W3DS a high bandwidth is needed, as, e.g., massive geometry and texture data must be transferred.

In contrast, the *OGC Web Perspective View Service* (WPVS) is a 3D portrayal service for generating and providing finally rendered *images* of perspective views of a 3D scene. A WPVS can provide high-quality visualization, regardless of a client's rendering capabilities, limited only by the rendering engine of the server. As only standard images are transferred, the WPVS can be used by more simple clients and in situations where available bandwidth is low. Additionally, server-side 3D rendering implies that we do not rely on a possible driver, outdated or even unknown client-side 3D hardware, i.e., we do not have to handle client hardware configurations. Using a WPVS, 2D views of complex 3DGeoVEs can be specified by simple URLs and can be easily integrated into various applications and systems such as web portals. Navigation in a base WPVS is step-by-step and far from real-time interaction. We argue that this drawback can be attenuated by intelligent loading and display strategies (e.g., image preloading).

A major difference between WPVS and W3DS is in the application of optimization strategies for rendering. In the case of W3DS-based portrayal, optimization strategies such as tiling, culling, caching, and level-of-detail mechanisms are implemented at the client side, which is in charge of selecting the right data to load from the W3DS. So, the W3DS client builds up a local context that is used for rendering. In contrast, with a WPVS optimization has to be implemented fully by the service, which does not provide any session management functionality. Thus, the WPVS rendering system cannot make any assumptions about the data to load and render. In worst case, an underlying (naive) rendering system would have to load and render different data for each single WPVS request.

2.2. Remote Visualization

Remote visualization comprises various approaches that distribute rendering functionalities within a network. In such systems, the data to visualize is processed remotely and transferred to a consumer as finally rendered image or video, or in a form that can be processed and rendered by the consumer more easily. Besides the OGC portrayal initiative, multiple other projects tackle the field of remote 3D visualization [9, 10].

At the client side, rendering techniques can be differentiated according to the type of data that is used as input for the rendering stage. They include, e.g., polygon-based rendering, point-based rendering, and image-based modeling and rendering. While both polygon-based and point-based rendering rely on geometry data, image-based rendering utilizes image information for deriving new views [3]. Image-based modeling includes techniques for reconstructing geometric information from images [5].

Mobile 3D visualization is mainly based on the programmable 3D graphics hardware that became available for mobile devices during the last years. It supports rendering 3D data directly on the mobile client. Here, a major challenge is to transfer geometry and texture data to the device. For this, elaborated data formats and rendering techniques for texture compression and buffer compression, frame buffer tiling, and geometry culling are required to reduce processing load and minimize power consumption [2].

2.3. Image Data and Formats

For rendering high-quality images, computer graphics hardware generates and takes into account per-vertex and per-fragment information such as vertex normal directions, texture coordinates, or depth values. This data is also incorporated by various rendering techniques for implementing specialized visual effects, e.g., object highlighting, edge enhancement, or ghost views. In computer graphics, the G-buffer [14] describes a concept for retrieving and storing this additional image information, which can be referred to as additional (i.e., non-color) image layers.

For enabling various visualization effects and techniques in post-processing, formats for storing these image layers are required. Especially for depth images, various work addresses data representation and transfer [7, 16]. OpenEXR [11] represents a format that supports several image channels in one image file, e.g., RGBA color values, luminance channels, depth values, and surface normals; it was originally designed for storing and editing high-dynamic-range (HDR) images.

Besides storing image layers as, e.g., arrays of float values, they can be encoded as color images. Storing data as images allows for utilizing standard image formats and their capabilities for data compression, which is crucial due to minimizing bandwidth usage when transferring image data over networks. Typical image compression algorithms include lossless Huffman encoding and Deflate. In [6], a technique that utilizes images as an efficient means for encoding geometry is presented. For each image layer type an encoding might exist that is more efficient than standard image encodings. However, in a standardization proposal in the context of service-oriented architectures, we rate more important the improved interoperability and reduced efforts for service consumers when using standard image encodings.

Various image formats allow for specifying the degree of compression, which leads to smaller data size, but possibly longer compression time and reduced image quality. Further, while lossy compression is acceptable for color values, it might be not for other image layers such as depth data.

3. Extending the Information Dimension

In this Section, we propose a concept for enhancing the WPVS by providing "semantic-rich" images, which complement the color images and serve as a basis for enhanced interaction and navigation.

3.1 Image Layers

Besides color images, we propose that the WPVS generates additional image layers storing information such as depth or object id values [14] for each pixel of the image. Fig. 1 shows examples of such image layers. We provide this additional data as separate images, in which each pixel value does not necessarily encode a color and is not necessarily dedicated for human cognition.

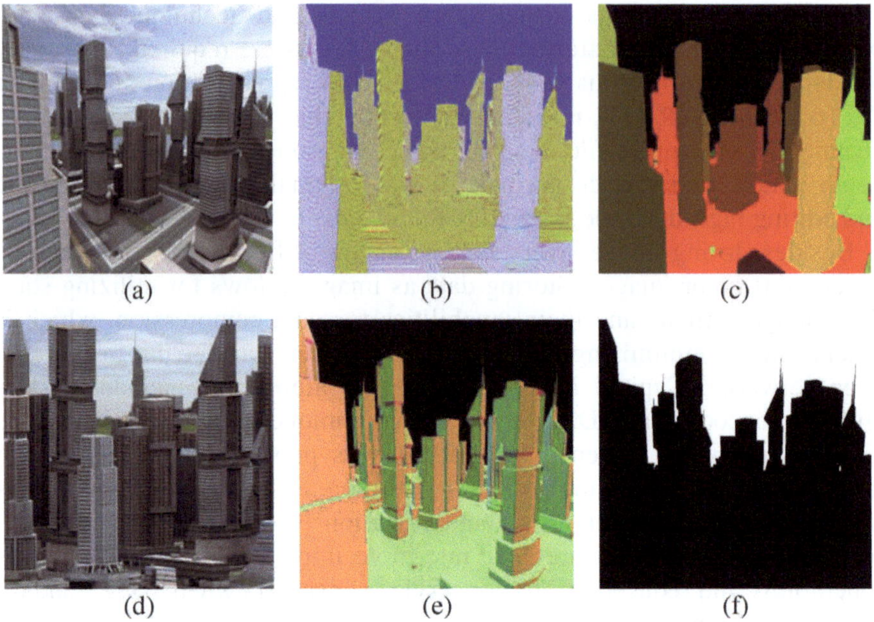

Fig. 1. Examples of image layers that are provided by *WPVS++*: color layer (a), depth layer (b), object id layer (c), normal layer (e), and mask layer (f). (d) shows an orthographic projection for the same location of the camera and point-of-interest.

Color Layer: A *color layer* contains a color value (e.g., RGB) for each pixel. According to the current WPVS specification, the service consumer can request a transparent background (RGBA), which requires image formats that support transparency, e.g., PNG or GIF.

Depth Layer: A *depth layer* describes the distance to the camera for each pixel. Depth images can serve for multiple effects, e.g., for computing 3D points at each pixel of the image (image-based modeling) or for various rendering techniques such as depth-of-field effects. Furthermore, depth information represents a major means to compose multiple images generated from the same camera position.

Typically, graphics hardware provides logarithmic and linearized depth values $z \in [0, 1]$. Different from this, we suggest providing depth values as distances to the virtual camera in world coordinates that correspond to the metrics defined by the requested *coordinate reference system* (CRS). This representation abstracts from computer graphical details and the distance values can be used without additional computation in many applications.

For this, each vertex p is transformed into camera space, and its linearized and normalized depth to the camera z_l is computed from its z-value p_z, and near plane *near* and far plane *far*. The graphics hardware interpolates the depth values for each fragment of an image; the value 0.0 is reserved for no-geometry fragments. In a second step for each image pixel, the depth value *depth* is computed in world metrics:

$$z_l = \frac{p_z - near}{far - near} \tag{1}$$

$$depth = near + z_l * (far - near) \tag{2}$$

Due to linearization, the precision of the depth values is decreasing for higher distances to the camera. For not reducing precision additionally, high bit rates are required for storing depth values.

As a first approach, we store depth values in IEEE 32 bit Float representation as images by segmenting their byte representations and assigning them to the color components of a pixel. Due to the direct storage as color components, resulting depth images possess very little pixel-to-pixel coherence, and should not be stored by lossy image formats. Thus, we suggest at least 32 bit (RGBA) images for this type of depth value storage.

Consequently, a service consumer has to recompose the depth values from the color components of the requested depth layer.

Object Id Layer: An *object id layer* contains a unique id for each pixel that refers to a scene object. Using this information, we can select all pixels that show a specific feature, e.g., for highlighting, contouring, or spatially analyzing the feature. For facilitating consistent interaction across multiple images, object ids should be unique over multiple requests. For that, they could be computed from an object id that is unique for the whole dataset.

Other Image Layers: In addition to the color, depth, and object id layers, we identified the following image layers to be relevant for the WPVS:

- *Normal Layer:* Describes the direction of the surface normal at each pixel. This could be used for the subsequent integration of additional light sources and the adjustment of the color values around that pixel, e.g., for highlighting scene objects. As a first approach, we encode normalized normal vectors in camera space by encoding each vector component (x, y, z) as color component (R, G, B) of a 24 bit color image.
- *Mask Layer:* Contains a value of 1 for each pixel that covers a scene object and 0 otherwise. Thus, a mask layer can be stored as 1 bit black and

white image. It could support the easy altering of the scene background or provide information about unused image space, which could be used, e.g., for blending advertisement, labels, etc.

- *Position Layer:* Contains the 3D coordinate at each 2D pixel. This allows the client to directly retrieve geo-position information, which, e.g., can be used for labeling or measurements.
- *Category Layer:* Classifies the image contents on a per-pixel basis; the classification could be based on, e.g., feature types, object usages, or object names. The category layer extends the so far described concept; it consists of an object id layer and a look-up table that lists for each object id the corresponding category (Fig. 2).

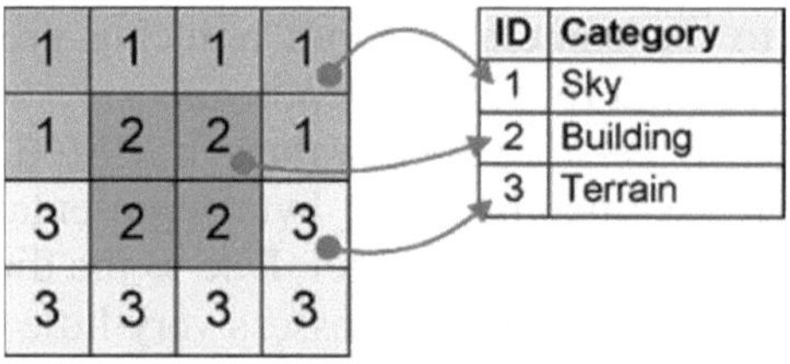

Fig. 2. A category layer consists of an object id layer and a category look-up table. Here, the pixels are categorized according to the feature type.

Image Data Encoding: Encoding also non-color data by standard image formats means to support the same principles for data encoding, data exchange, and client-side data loading and processing for all image data. Additionally, using images allows for applying state-of-the-art compression algorithms. For color data, which is dedicated for perception by humans, lossy compression algorithms can be used; for this data, JPEG mostly represents a suitable data encoding format. For data that requires exact values for each pixel, lossless image compression must be applied. We suggest PNG to encode this image data. PNG is also used for color images that shall contain transparency.

Technically we encode JPEG images by 24 bit per pixel and PNG images by 24 or 32 bit per pixel. Higher data accuracies could be reached by using image formats with higher bit rates, e.g., 32 (48) bit images or even 48 (64) bit images. Drawbacks include increased image size and, thus, transfer load, and processing time at both the service-side and the consumer-side.

3.2. Image Sets

Various applications could require to fetch multiple, spatially coherent image layers from the *WPVS++*. As the WPVS is a stateless service (i.e., each service request is self-contained and its processing is independent from any preceding request), it potentially has to fetch, map, and render disjunctive sets of geodata for each service request. In particular, this is relevant when considering large 3DGeoVEs or remote data sources to be visualized.

From a performance point of view, it is better to support multiple image layers within a single *WPVS++* response. Provided that the requested views are spatially coherent (i.e., they are located close to each or have overlapping view frustums), this could reduce the number of switches of the rendering context. As an additional benefit, the network communication overhead caused by multiple requests is reduced.

These image sets could provide multiple images layers for the same camera specification (e.g., a set consisting of a color layer, depth layer, and object id layer) as well as images for different camera specifications (e.g., perspective views along a path through the 3D world).

3.3. Convenient Camera Specification

The current OGC WPVS provides a camera specification that is based on a point-of-interest (POI) and is not convenient for many applications. For example, specifying a setting that shows a view out of a window to a specific point in the 3D world demands for laborious computations for deriving distance, pitch, yaw, and roll angles.

As various applications could profit from a different specification of the view frustum, we suggest replacing the camera specification by another one that is based on only three vectors: the 3D coordinate of the camera position (`POC`), the 3D coordinate of the point-of-interest (`POI`), and an optional camera up vector (`UP`) for specifying camera rolls (List. 1). Furthermore, parameters for near plane and far plane are required; they describe the culling volume used during the rendering process and are necessary, e.g., for the generation and usage of depth image layers. For the specification of distorted images, we suggest replacing the angle-of-view parameter by a field-of-view angle in degrees in horizontal direction (`FOVY`) and to complement it by the optional vertical angle (`FOVX`). If `FOVX` is not specified, it is derived from the image dimensions.

3.4 Additional Projection Types

The current OGC WPVS is intended for central perspective projections only. Complementary, an image-based 3D portrayal service could offer additional projection types, such as orthographic projections, which are used, e.g., for architectural applications.

For extending the *WPVS++* for orthographic projections, we suggest an alternative camera specification `Orthographic` that is controlled by the parameters `Left`, `Right`, `Top`, and `Bottom`, which describe the borders of the cuboid view frustum (List. 1).

Beyond perspective and orthographic projections, a 3D portrayal service could support more advanced projection types such as panoramic views (i.e., field-of-view larger than 180 degrees) or multi-perspective views [8].[2] While some of these projections could be generated from 2D images as provided by the current WPVS implementation, others require geometric information and, thus, are an integral part of the rendering process and need to be implemented by the 3D portrayal service itself.

3.5. Extended `GetView` Operation

A *WPVS++* `GetView` request mainly specifies the datasets to include, visualization styles to apply, and a camera specification; this information can be called image context. Adding the concept of multiple image layers, a *WPVS++* image context could be specified as described in List. 1.

Style specifications within this image context are dedicated and applicable only for requested color images. Parameters for the requested image layers allow for specifying their contents, e.g., a category image could be parameterized with the category to apply (e.g., feature type or building usage) and mask images could be generated for specific features (e.g., specified by GML-id). Using this image context definition, the *WPVS++* `GetView` operation is modified and extended in respect of multiple camera specifications as described in Listing 1.

4. Extending the Interaction Dimension

We consider the support of interaction with the generated image as a main requirement of the *WPVS++*. This includes techniques for maneuvering

[2] Due to possible non-perspective projections, the name of the *WPVS++* should be adapted in future, e.g., to Web View Service.

the virtual camera through the 3D world and for retrieving information about the visualized objects. In this section, we describe extensions of the WPVS that support navigation within and information retrieval from the rendered images.

Listing 1. Extended interface of the *WPVS++*.

```
OP GetView (mand.)
 IN
  ImageContext(1,n)
 OUT
  Image(1,n): Image

OP GetCategory (opt.)
 IN
  CategoryType: string
  ObjectId: integer
 OUT
  Category: string

OP GetPosition (opt.)
 IN
  ImageContext: ImageContext
  Coordinate2D(1,n): Vector2D
 OUT
  Coordinate3D(0,n): Vector3D

OP GetFeatureInfo (opt.)
 IN
  ImageContext: ImageContext
  Coordinate2D(1,n): Vector2D
  ResponseFormat: string
 OUT
  FeatureInfo(0,n): anyType

OP GetNavigationCamera (opt.)
 IN
  ImageContext: ImageContext
  Coordinate2D(1,n): Vector2D
  NavigationType: MoveCamera
  ResponseFormat: string
 OUT
  Camera : Camera

OP GetMeasurementResult (opt.)
 IN
  ImageContext: ImageContext
  Coordinate2D(2,*): Vector2D
  MeasurementType: PATH | AREA
  ResponseFormat: string
 OUT
  MeasurementResult: string

TYPE Camera
 PointOfCamera: Vector3D
 PointOfInterest: Vector3D
 UpVector: decimal (opt.)
 Proj: Perspective|Orthographic
 NearPlane: decimal
 FarPlane: decimal

TYPE Perspective
 FovY: decimal
 FovX: decimal (opt.)

TYPE Orthographic
 Left: decimal
 Right: decimal
 Top: decimal
 Bottom: decimal

TYPE ImageContext
 CRS: string
 Dataset(1,k): Dataset
 BoundingBox(0,1): BoundingBox
 Style(1,k) : string
 SLD(0,1): SLD-Ref
 Camera(1,m): Camera
 Width: integer
 Height: integer
 ImageLayer(1,n): ImageType
 OutputParameter(0,n): string
 OutputFormat(1,n): string

ENUM ImageType
 { COLOR, DEPTH, OBJECTID,
   NORMAL, MASK }
```

As described, feature-related information could be bulk-loaded to a *WPVS++* consumer as non-color layers, e.g., object id layers or category layers. This is particularly useful, when this data is required for a lot of pixels of the image, as bulk loading avoids the transmission overhead caused by multiple requests. However, if only sparse information is needed, or in the case of a very simple client, an operation for requesting particular feature information at a 2D pixel position would be suitable; transfer load would be reduced and a client would not have to parse and evaluate the whole image layers.

4.1. Requesting 3D Coordinates

Retrieving 3D geocoordinates for a specific 2D pixel position in a generated image is useful for many applications (e.g., for specifying the position of a defect fireplug). For this, we propose a *WPVS++* operation `GetPosition` (Listing 1). As the *WPVS++* is stateless, in addition to the 2D pixel position the image specification of the original `GetView` request has to be part of the `GetPosition` request. This operation is highly useful for navigation in the visualized 3DGeoVE: A user could select two positions in the image that specify the desired point of camera and point of interest; the client requests corresponding 3D coordinates, and uses these for specifying the camera in a subsequent `GetView` request.

4.2. Requesting Feature Information

Similar to the WMS, a `GetFeatureInfo` operation should provide extra information about the geoobjects at a specific pixel position. This operation could either provide domain-specific or any potentially useful information to the consumer. According to the underlying data source and its capabilities, various response formats could be supported, e.g., attribute names and values in XML format or even GML structured data sets. Additional formats can be supported and specified as mime type within the service request. In the case of structured responses (e.g., in GML format), the WPVS operation `GetDescription` can be used for retrieving schema information for specific datasets.

4.3 Navigation based on Picking

Navigation represents a fundamental interaction type for 3DGeoVEs. From virtual globes and online map systems users know to navigate by using mouse and keyboard in a real-time interactive manner. However, especially due to the image-based approach, the WPVS inherently provides only non-real-time step-by-step navigation. Assistant navigation techniques could compensate that drawback and allow a user to interact with the image, e.g., by sketching a desired point of interest. The portrayal service automatically interprets this navigation input, taking into account scene objects, their types, and their navigation affordances, and computes and responds a camera specification, which can be used for requesting the corresponding view.

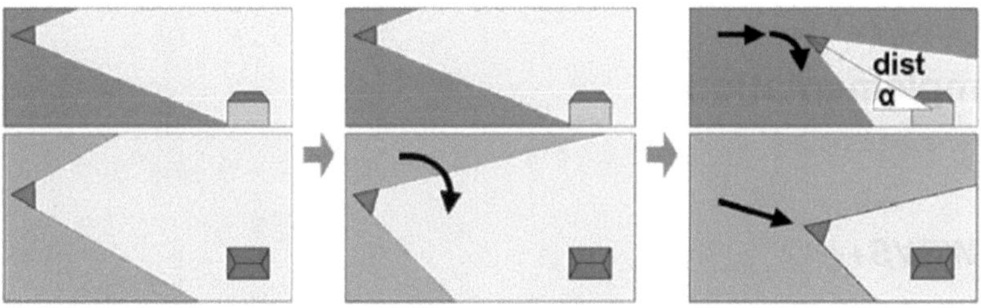

Fig. 3. Changing the camera orientation and position; the camera is oriented to the selected object, moved there while keeping its height, and finally looks down in a specified angle. Side view (top) and top view (bottom).

However, complex, arbitrary, or domain-specific navigation techniques can hardly be defined and described in a formal interface specification. Because of that, we propose an optional generic operation `GetNavigationCamera` that can compute and provide a meaningful camera specification for a 2D pixel position (List. 1). For virtual 3D city models, we suggest a comprehensible navigation technique that supports a user in moving closer to a selected object. This technique moves the camera toward the center of the bounding box of a selected object, while the camera height is kept constant; finally, the camera is positioned in a specified angle above the object (Fig. 3). This ensures that the camera is only moved along a line of sight keeping the probability of occlusion low.

4.4. Measurement based on Picking

Measuring within a requested view represents a main functionality used to retrieve information about the spatial extent of geofeatures, their spatial relationships, and the overall spatial layout of a 3D scene. Hence, *WPVS++* should support the measurement of distances, paths, and areas (List. 1):

- *Path measurement:* Computes the sum of the Euclidean distances between the 3D positions derived from each pair of consecutive 2D pixel positions. If only two 2D pixel positions are provided as input, the *WPVS++* performs distance measurement.

- *Area measurement:* Computes the area which is outlined by the input points. In 3D, area measurement is not straight forward, as the derived 3D points are not likely to be coplanar. Projecting the 3D points onto a horizontal 2D plane would be a straight forward solution for this problem. More accurate algorithms are possible, e.g., computing a best fitting plane for projection or applying differential techniques to the problem.

5. Implementation

5.1. *WPVS++*

For demonstrating the applicability of the proposed concepts and extensions, we have implemented a prototype of the described *WPVS++* service and tested it with a virtual 3D city model. This implementation is based on the Virtual Rendering System (VRS). The demo dataset is given in 3DS format. Thus, only object names and unique object ids can be requested via `GetFeatureInfo` operation. The service is accessed by an HTTP interface on top of TCP/IP.

For showing the applicability of the *WPVS++*, we have implemented two client applications, which differ in their requirements concerning processing power, rendering capabilities, and degree of interactivity.

5.2. Lightweight JavaScript Client

As example of very lightweight clients, we implemented a JavaScript-based client, which can be run in a web browser without additional plug-ins or libraries. This client just requests and displays color images by HTML `<img/>` elements. It provides step-by-step navigation such as moving left/right, tilting up/down, or rotating around the POC or POI. For each navigation step a new image is requested and displayed.

Moreover, this client takes advantage of the `GetPosition` operation for retrieving 3D coordinates for changing the virtual camera's orientation and for implementing a "move there and orient here" navigation. The `GetFeatureInfo` operation is used for retrieving object information about features in the image.

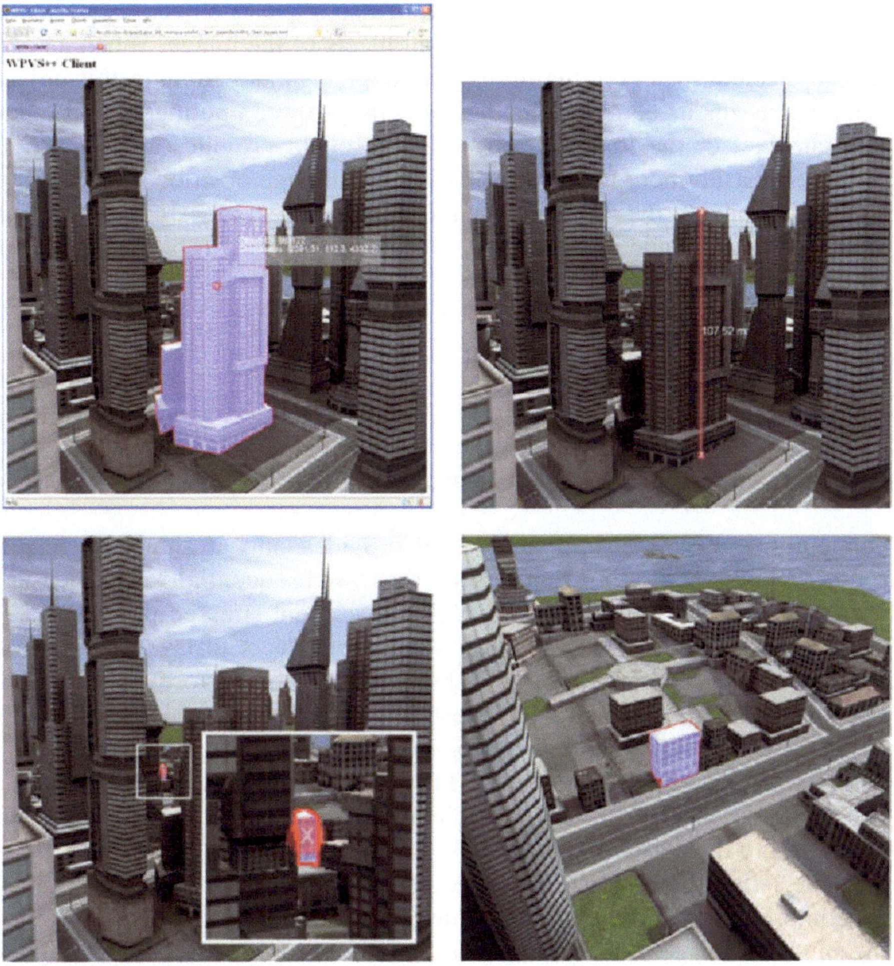

Fig. 4. Java-based interactive web client. Object selection by using the object id layer (upper left); distance measurement by calculating 3D positions from the depth layer (upper right); moving the camera to a selected object by requesting *GetNavigationCamera* (lower left and right).

5.3. Java-based Client

A second client application (Fig. 4) implements functionalities for image processing and rendering; thus, it requires more processing power. It is Java-based and can be integrated into a web page as a Java applet, which also allows for a broad range of applications.

In addition to the color layer, this client also requests object id and depth layers. Object ids are used for selecting objects of the image. When an object is selected, the client highlights all the pixels that belong to this object and emphasizes its contour. Depth data is used for computing the 3D point at each pixel of the image. Using this, the client provides an interactive distance measurement tool. After selecting an image position, the distance from this point to the 3D point under the mouse cursor is computed and displayed as an image overlay.

6. Results and Discussion

6.1. Measuring the *WPVS++* Response Behavior

Our current *WPVS++* implementation provides only a single rendering server. Thus, synchronous requests are processed sequentially. For measuring the service's response behavior, the Java client stressed the *WPVS++* by repeatedly requesting 120 color, depth, and object id layers. Up to 10 requests have been sent in parallel. Furthermore, the performance of multipart responses was tested by requesting 40 image sets (color, depth, and object id layer). The test was performed in an intranet environment; rendering was performed on a desktop PC (2.4GHz double core, 2 GB RAM, NV GeForce 7900 GS), the Java client was executed on a different PC connected to the network.

Table 1 lists the number of images per second that the service can render, compress and send, and the client can decompress. Generally, providing multiple images in one response via HTTP multipart provides higher throughput than requesting each image layer separately. This could be caused by reduced message overhead, i.e., sending and processing the requests. The gain increases for smaller images.

Table 1. Maximum number of processed requests per second when requesting color layer, depth layer, and object id layer separately (single part response) or within a single request (multipart response) using three image sizes.

	256 x 256	512 x 512	1024 x 1024
Single part response	4.7	3.0	1.5
Multipart response	13.9	5.7	2.6
Factor	2.95	1.90	1.73

6.2 Measuring the Size of Image Layers

We logged the size of generated and transferred image layers while navigating through a virtual 3D city model at the Java client. For an overall number of 370 camera positions, the client requested color layers (24 bit JPEG), depth layers (32 bit PNG), and object id and normal layers (24 bit PNG each) in a size of 512 x 512 pixels. The navigation process included overviews as well as pedestrian views. While the depth layer accounts for nearly two-thirds of the transmitted data, object id layer and normal layer play only a minor role (Table 2). This is because for depth layers, only low compression rates can be reached, due to the little pixel-to-pixel coherence. Thus, alternative encodings for mapping depth values to image data are required in order to reach better compression rates.

Table 2. Image layer size in Kbytes when navigating through a virtual 3D city model and percentage of each image layer type.

	Minimum	Maximum	Average	Avg. percentage
COLOR	8.37	126.11	77.95	0.25
DEPTH	5.98	725.22	199.63	0.64
OBJECTID	3.50	19.05	9.56	0.03
NORMAL	3.50	85.95	24.30	0.08

6.3. Comparison to full 3D Geovirtual Environments

Compared to full 3DGeoVEs, the *WPVS++* provides only visual representations. Modalities such as auditory or tactile perception, are not addressed. However, provided a sufficient bandwidth, we assume high usability rates for applications and systems based on the *WPVS++*.

Navigation capabilities mainly depend on the functionality of the clients that consume the *WPVS++*. Simple clients can provide a step-by-step navigation based on requesting and displaying single views. For this, the additional `GetPosition` operation facilitates an easy and targeted camera manipulation. Based on depth information, more complex clients could apply, e.g., image-based rendering techniques for providing more convenient and close to real-time visualization and navigation functionality.

Regarding object interaction, the *WPVS++* supports retrieving information for individual objects of the scene. Object manipulation capabilities are limited: For persistent manipulations of feature geometries and attributes the original data sources must be accessible, e.g., via a Web Feature Service (WFS) [17]. For user-defined appearance manipulation, effective styling mechanisms and specifications are required.

7. Conclusions and Future Work

In this work, we present and discuss the *WPVS++*, an extended version of the OGC Web Perspective View Service (WPVS), which aims at enhanced interaction and navigation capabilities. The *WPVS++* includes functionalities for the exploration and analysis of presented 3DGeoVEs by simple clients, e.g., by requesting 3D positions, retrieving object ids and measuring distances. Additionally, the *WPVS++* provides non-color image layers, which represent thematic information and can be utilized for enhancing client-side interaction. We demonstrate the implementation of major *WPVS++* features with two different client applications and a common application example.

The extended *WPVS++* contributes to making complex and massive 3DGeoVEs accessible and interactively usable. Due to the HTTP-based interface, it can be easily integrated into web-based applications and systems, e.g., for urban planning, public participation systems, location marketing, or threat or emergency response. The additional non-color image layers of the *WPVS++* facilitate advanced image processing capabilities within spatial data infrastructures, e.g., for image-based annotation of the 3DGeoVE or for advanced rendering effects. Within a spatial data infrastructure, the *WPVS++* could serve as frontend to a Web 3D Service (W3DS): The *WPVS++* could request the 3D data as scene graph from W3DS, and render and distribute images.

In future work, we plan to fully implement the proposed functionalities, to further formalize the extension concepts, and to research techniques for further improving the interaction process in terms of efficiency and interactivity. For example, we will investigate how to better compress depth image layers while still using standard encodings. As a further task, we will extend the client application by image-based rendering techniques for better using the potentials of the image-based portrayal service. Additionally, we will investigate the service-based provisioning of non-photorealistic rendering and illustrative depictions, which could support and ease the cognition of 3D views.

References

1. Altmaier A, Kolbe Th (2003) Applications and Solutions for Interoperable 3D Geo-Visualization. In: Fritsch D (ed) Photogrammetric Week. Wichmann
2. Capin T, Pulli K, Akenine-Möller Th (2008) The State of the Art in Mobile Graphics Research. Computer Graphics Application 28(4):74–84

3. Chang C and Ger S (2002) Enhancing 3D Graphics on Mobile Devices by Image-Based Rendering. In: Proc. of the 3rd Pacific Rim Conference on Multimedia (PCM '02), Springer, pp 1105–1111
4. de la Beaujardiere J (2006) OpenGIS Web Map Server Implementation Specification. Version 1.3.0, Open Geospatial Consortium Inc.
5. Faugeras O, Laveau S, Robert L, Csurka G, Zeller C (1995) 3D Reconstruction of Urban Scenes from Sequences of Images. In: Gruen A, Kuebler O, Agouris P (eds) Automatic Extraction of Man-Made Objects from Aerial and Space Images. . Birkhuser, pp 145–168
6. Gu X, Gortler SJ, Hoppe H (2002) Geometry images. ACM Trans. Graph 21(3):355–361
7. Lapidous E, Jiao G (1999) Optimal depth buffer for low-cost graphics hardware. In: Proc. of the ACM SIGGRAPH/EUROGRAPHICS workshop on Graphics hardware (HWWS '99). New York, ACM, pp 67–73
8. Lorenz H, Trapp M, Jobst M, Döllner J (2008) Interactive Multi-Perspective Views of Virtual 3D Landscape and City Models. In: Proc. of the 11th AGILE Int. Conf. on GI Science. Springer
9. Nadalutti D, Chittaro L, Buttussi F (2006) Rendering of X3D content on mobile devices with OpenGL ES. In: Proc. of the 11th Int. Conf. on 3D web technology (Web3D '06). New York, pp 19–26
10. Nurminen A (2008) Mobile 3D City Maps. In: IEEE Computer Graphics and Applications 28(4):20–31
11. OpenEXR documentation. http://www.openexr.com/documentation.html (14.06.2009).
12. Percivall G (ed) (2008) OGC Reference Model. Version 2.0, Open Geospatial Consortium Inc.
13. Quadt U, Kolbe Th (ed) (2005) Web 3D Service, Version 0.3.0, OGC Discussion Paper, Open Geospatial Consortium Inc.
14. Saito T, Takahashi T (1990) Comprehensible rendering of 3-D shapes. In: SIGGRAPH Computer Graphics 24(4):197–206
15. Singh RR (ed.) (2001) Web Terrain Server (WTS). Version 0.3.2, OGC Discussion Paper, Open Geospatial Consortium Inc.
16. Verlani P, Goswami A, Narayanan PJ, Dwivedi S, Penta SK (2006). Depth Images: Representations and Real-Time Rendering. In: Proc. of Int. Symp. on 3D Data Processing Visualization and Transmission, pp 962–969
17. Vretanos PA (ed) (2005) Web Feature Service Implementation Specification. Version 1.1.0, Open Geospatial Consortium Inc.

Interactive modelling of buildings in Google Earth: A 3D tool for Urban Planning

Umit Isikdag and Sisi Zlatanova

Abstract. Urban planning and renewal is a very complex process consisting of tasks that require joint decision making. In such tasks, the communication tools to present and convey design ideas are of critical importance. Appropriate visual tools are needed to establish a common language between professionals, and between professionals and citizens. However, there are indications that particular representations and tools may create different perceptions with respect to the background of the actors. Therefore it is important to develop simple tools which will allow specialists and non-specialists to demonstrate their ideas during the discussions. New emerging technologies offer a large variety of visualisation tools that may solve this problem and greatly enrich visualisation and communication possibilities.

This paper presents an approach to draw and visualise simple geometric representation of buildings directly in the Google Earth environment. Using this tool the urban planners and citizens can introduce proposed buildings into the existing 3D environments in a commonly used application such as Google Earth. Several alternatives for the proposed buildings can be easily created, examined and evaluated (in the virtual environment). Using the proposed tool the citizens can generate their own solutions for a design evaluation scenario at home or during a joint decision making session at a municipality and present them to the specialists. The developed tool is Java-based and uses newly released API of Google Earth.

T. Neutens, P. De Maeyer (eds.), *Developments in 3D Geo-Information Sciences*,
Lecture Notes in Geoinformation and Cartography, DOI 10.1007/978-3-642-04791-6_4,

Introduction

Urban planning and renewal is practically a complex decision-making process, which involves many actors, who perform different tasks and have diverse points of view. In such projects, the actors have often-conflicting expectations, background and foci. The steps to approve a planning (or design) proposal or a renewal project are many, as the approach varies per municipality and country. For example, in Netherlands six phases can be distinguished starting from a very general vision for development and finishing with a detailed architectural plan of each building in the area (Zlatanova *et al* 2008, Kibria *et al* 2009). Actors involved in those phases are quite different. As the phases become specific the actors change from urban planners and local authorities to architects and citizens (especially in the last three phases when the volumetric design is presented). Recently completed studies (e.g. Kibria, 2008) have clearly shown that these phases also require the use of 3D representations and visualisations to support the joint decision making process during the evaluation of design / renewal proposals. As the current state of the art municipalities use 3D physical (wooden, plastic or carton) models to discuss the projects. The representations of the buildings are very often schematic (i.e. in form of simple boxes) and the citizens are encouraged to play with them and develop new volumetric compositions (by using these physical models), to develop and convey an agreed point of view.

Traditionally, during evaluation of design / renewal proposals municipalities use 2D paper maps, CAD drawings, graphical images, textual/oral information and 3D physical models to present ideas and discuss alternatives to citizens. Most municipalities have websites for dissemination of spatial plans, but they are in 2D and in most cases offer only static visualisation. Only recently large municipalities may employ systems allowing for presenting interactive digital maps for the public in Web Map Services (WMS) (Knap and Coors, 2008). In the last several years, advances of geo-information technology such as 3D virtual environments, 3D analytical visualisation and 3D formats for sharing of data, offer large spectrum of new possibilities for communication of ideas and discussion of design alternatives. Virtual environments like Google Earth, Virtual Earth have made 3D visualisation of urban fabric known and accessible for everyone.

Many municipalities and urban designers are in the process of discovering these new functionalities. However, there are indications from practitioners that this visual information is not necessarily equally perceived. A given representation may be misunderstood by the receiver, due to a wrong presented message or due to different concepts of the actors. Therefore, municipalities are looking for new tools that allow actors to actively interact with the design to enable them express their own understanding about size and height of a building, location and direction. Despite recent developed tools and systems to support 3D visualisation for municipalities (e.g. Kolar *et al* 2008, Reitz *et al* 2009) and virtual exploration of cities (e.g. www.cebra.eu), the interest in virtual globes (3D Geospatial Browsers) as Google Earth is increasing. Being freely available, citizens are largely familiar with the navigation and interaction options in these tools and can participate relatively easily in the discussions. A growing number of municipalities use Google Earth to announce events, renewal projects or just to provide a 3D view of the city.

All these developments encouraged us to look for tools that will allow for a level of communication and interaction between citizens and local governments, which suits exchange of ideas in virtual worlds. The aim of the ongoing research presented in this chapter is:

"providing an easily accessible visualisation tool to support the joint decision making process (by facilitating the user interaction in modeling) during the evaluation of a design/renewal proposal"

The research mainly concentrates on facilitating the interactivity in generation of 3D visualisation of buildings. Although this tool will facilitate the joint decision making process during a design evaluation session held in municipalities, the authors would like to clarify that this research is uniquely focused on the enabling and facilitating interactive visualisation (of buildings). In addition, it should be taken into account that the presented research is not focusing on developing a software (or a tool) that would support the overall public participation process.

The research and literature on developing tools for enabling and facilitating planning support and public participation is extensive. The following provides only a few notable examples of the research (and development) from the literature in the field, which the reader can refer. As an experienced researcher in the area, Al-Kodmany (2002) reviewed both traditional and computerized visualization tools for public participation and provided a general map for planners as they navigate through the multitude of options that exist for visualization in public participation planning. Hudson-Smith *et al.* (2002) presented a UCL CASA working paper based on a UK case study which is focused on the impact of digital communication and online participation, where the users also made use of 3D visual representation of the urban elements. For several years, Environmental Simulation Center has been carrying out studies on 3D visualisation and participation in projects such as Baltimore Vision 2030, Lower Manhattan,

Santa Fe (ESC,2009). A recent EU project VEPS, -Virtual Environment Planning- (Knapp and Coors, 2008) was focused on developing tools for facilitating the communication in the planning process with 3D visualisation and efficient use of the web technologies. There is also commercial software such as CommunityViz (CommunityViz, 2009) which support the public participation process by 3D visualisation. In general terms, ongoing research on facilitating the public participation by the use of ICTs and software tools developed for this purpose focus on a number of issues including:

- Visualisation of the urban environment
- Visualisation of design proposals
- Online communication of users
- Gathering user feedback
- Analysing the social and environmental impact of proposals

Conversely, the tool that is being developed during this research can not be classified as one of these tools (i.e., a public participation enabler or a facilitator) as it only focuses on a single aspect of the overall process, but it can only be regarded as a medium for facilitating the rapid 3D visualisation of design proposals (by user interaction) in joint decision making sessions (held physically or virtually).As the tool is intended for quickly presenting ideas in a well recognised 3D Geospatial Browser (Google Earth) it can be used as a communication/visualisation facilitator for different actors in urban development and renewal process.

The chapter is organised as follows: following the introduction section which covers the aim of the research, the second section briefly outlines the role of Google Earth as the new geospatial medium of visualisation and communication, and summarises the representation methods of building models within Google Earth. The third section illustrates the approach for preparing digital models of existing buildings (which will act as the background data for the design proposal evaluation process). The fourth section introduces a new method for generation of building geometries based on the user interactions on Google Earth, and provides the details of the prototype implementation realized by using this method. The chapter finalizes with the conclusion section, which also provides the future research directions.

Google Earth as tool for visualisation and communication

3D Geospatial Browser is a new term for defining browsers that are capable of retrieving geospatial data over the web and visualising it over a virtual globe (in form of raster and vector representations). Most known 3D Geospatial Browsers are Google Earth, Microsoft Virtual Earth and NASA World Wind. The data that forms the base layer of the virtual globe is satellite images with varying resolution where the resolution of the images change depending on the user interaction with the browser. In most 3D Geospatial Browsers it is possible to search for an address, a point of interest, as well as a coordinate pair. In addition, it is also possible to navigate (pan, zoom, rotate etc.) by using simple and user-friendly mouse functions or appropriate toolbars.

Google Earth is the most popular 3D Geospatial Browser. In Google Earth, for most parts of the world the terrain data is available in form of the Digital Elevation Model (DEM), and the terrain can be visualised in 3D, although for many areas not that accurate. The user input is a very important resource of data, and there are many map layers created by the user input. Users can create their own points and areas of interest (i.e. placemarks), define their geometries and associate semantic information with these points and areas of interest.

Fig. 1. Visualisation of two 3D building models in Google Earth

Google Earth has the ability to visualise the 3D representations (models) of the buildings which are created by the user input. These models can then be viewed by all the users of the software. The buildings visualised in Google Earth, are usually created through a 3D modelling software such as SketchUp or other 3D modelling software and imported as KML/KMZ file formats. Figure 1 shows 3D visualisations of two buildings within Google Earth.

The representation of the building in Google Earth is a BRep (Boundary Representation) model, in addition, these models can also have textures associated with them. Google Earth was originated from the software called Earth Viewer, created by Keyhole Inc. Thus, the mark-up language (used to represent the geometric models in Google Earth) is known as Keyhole Markup Language (KML). KML supports representation of objects by using the BRep method. The language is based on the XML standard and uses a tag-based structure with nested elements and attributes. In KML, the building geometries are represented with polygons. Two distinct methods exist to represent the building geometries. First one is, defining a base polygon that corresponds to the floor plan of the building and extruding this base polygon to the height of the building. Figure 2 illustrates 4 buildings represented by this method.

Fig. 2. Representation of 4 buildings within base polygon extrusion (Google Code, 2009)

The figure above presents the visual representation of 4 buildings of the Google Campus. Their representation method is explained in the KML documentation (Google Code, 2009). The building located top-right of the illustration is named as Building 40, and represented by a set of semi-transparent red polygons. The polygons representing the walls and the roof of the building are generated on-the-fly (i.e. at run time) by Google Earth, based on the coordinates of the base-surface and the extrusion parameter.

The KML code to represent Building 40 in Google Earth is provided in Table 1.We refer this type of representation as *Base Polygon Extrusion.*

Table 1: The KML code used to represent the Building 40 (Base Polygon Extrusion)

```
<Placemark>
    <name>Building 40</name>
    <visibility>0</visibility>
    <styleUrl>#transRedPoly</styleUrl>
    <Polygon>
     <extrude>1</extrude>
     <altitudeMode>relativeToGround</altitudeMode>
     <outerBoundaryIs>
      <LinearRing>
       <coordinates>
        -122.0848938459612,37.42257124044786,17
        ...
        -122.0848938459612,37.42257124044786,17
       </coordinates>
      </LinearRing>
     </outerBoundaryIs>
    </Polygon>
   </Placemark>
```

The second method for representing the building's geometry is, using multiple polygons. This method enables more detailed geometrical representation of the building elements and is currently used by several different applications that acquire/transfer information from digital building models into Google Earth. In this case, the 3D Geospatial Browser will visualise every face of the building -one-by-one, based on the parameters provided in the KML code. We refer this type of representation as *Multi Polygon Representation.* An extract from the KML code for representing the same building with multiple-polygons is provided in Table 2.

The users will not notice the difference between both representations, when a building's geometry is represented in a low level of detail (i.e. Figure 2, in LODs similar to CityGML LOD1, www.citygml.org). The first representation is beneficial for large data sets, where whole cities have to be visualised. For example, tests performed with all the buildings of the city of Delft have shown drastic reduction of the time needed to load KML file when the buildings are represented using the base extrusion approach. However, it would become impossible to represent the geometry of building elements by using the first method, if a detailed representation / visualisation of, the building façade and indoor building elements is required (i.e. in LODs similar to CityGML LOD 3, LOD 4), and, in that situation the *Multi Polygon Representation* becomes an inevitable need.

Table 2: The KML code to represent the Building 40
(Multi Polygon Representation)

```
<Placemark>
    <name>Building 40</name>
    <visibility>0</visibility>
    <styleUrl>#transRedPoly</styleUrl>
    <Polygon>
     <extrude>0</extrude>
     <altitudeMode>Absolute</altitudeMode>
     <outerBoundaryIs>
      <LinearRing>
       <coordinates>
        ...
        </coordinates>
      </LinearRing>
     </outerBoundaryIs>
    </Polygon>
    <Polygon>
     <extrude>0</extrude>
     <altitudeMode>Absolute</altitudeMode>
     <outerBoundaryIs>
      <LinearRing>
       <coordinates>
        ...
        </coordinates>
      </LinearRing>
     </outerBoundaryIs>
    </Polygon>
     ....
     ....
 </Placemark>
```

Generation of existing buildings

To be able to present and discuss new 3D design alternatives, first a 3D model of the existing area should be available. Three dimensional representation of urban environment has been an active research topic in last years. Various approaches exist to reconstruct 3D models, where information about an existing building is collected from (a) single or multiple source(s) and geospatial models are created with respect to an application (Tao, 2006). Among all the approaches, 3D Laser Scanning Technology has emerged as the most innovative method and much research is devoted to developing automatic algorithms for 3D reconstruction (i.e. Arayici,

2007, Kang *et al* 2007, Pu 2007). In fact, such approaches for gathering geometric information about building façades are commonly used for important public or historical buildings. However, in many cases urban planners do not require such detailed 3D models for all buildings. The existing situation is mostly used as background information to demonstrate new ideas. Therefore the 3D models used in urban planning can be relatively simple.

In this respect, a very appropriate 3D re-construction approach for a municipality considers a combination of existing 2D digital maps and LiDAR data. The 2D digital maps provide the outlines of the buildings and the LiDAR data are used to derive height information per building. The buildings are then extruded to the derived height. Depending on the purpose, these 3D blocks can be further decorated with actual textures (photos) to increase the realism, but texturing is not applied in most of the cases. Municipalities usually have such data at their disposal and the simple buildings can be created in an automatic manner. Almost all commercial packages allow the generation of 3D extrusion (from given polygons and a height value per polygon).

An approach for 3D reconstruction and export in KML is presented in Kibria (2008) and tested for the area of Delft University of Technology. The approach consists of several steps: 1) import of the 2D digital map and the LiDAR data, 2) filtering all laser points that are outside the outlines of the polygons, 3) selecting the mean value for the height of each building, 4) extruding the 2D polygon (outlining the building) with the given height, and 5) exporting the created buildings to KML with the help of Arc2Earth extension. The resulting KML file is structured according to the first method (as specified in Section 2). Figure 3 illustrates some of the steps.

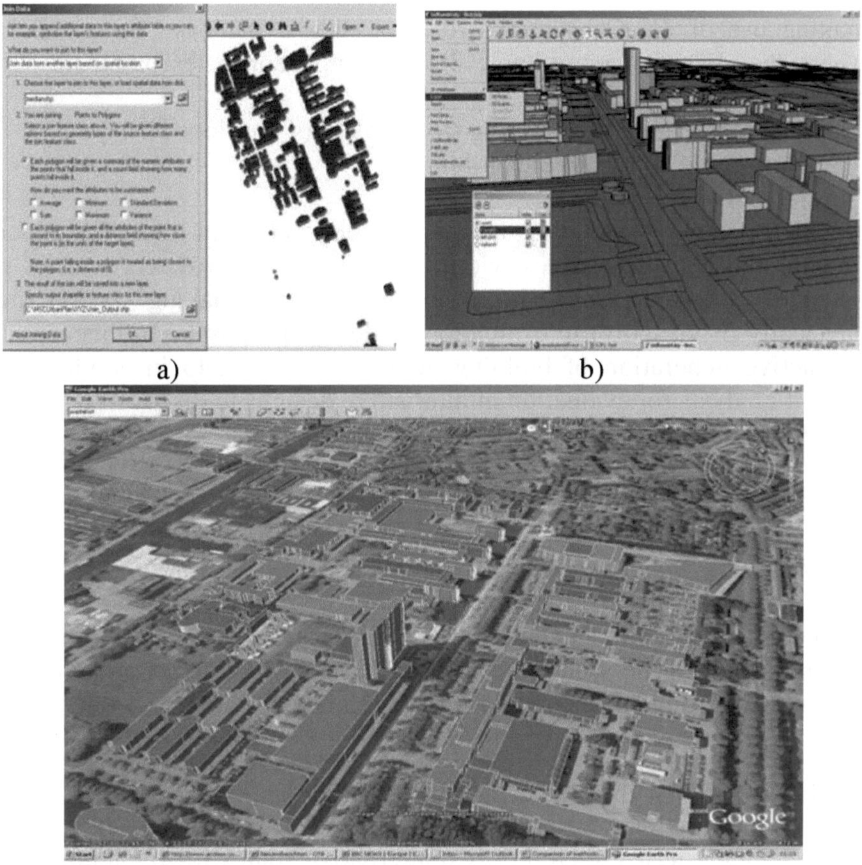

a) b) c)

Fig. 3. Snapshots of a) filtering laser scan points in the outlines of the buildings, b) 3D reconstructed model in ArcGIS and c) 3D block model of buildings in Google Earth. (courtesy, Kibria, 2008).

It should be noted that the same result can be obtained using AutoCAD (AutoDesk), FME (Safe software), Bentley software, etc. All the packages provide tools to create scripts which allow for partial or complete automation of this process.

Generation of buildings directly in Google Earth

In an IT supported joint decision making session which is held by municipalities it should be possible for the stakeholders to draw the alternative design proposals in the virtual environment. The idea behind the approach presented here is providing an interface to the stakeholders, by making use of 3D Geospatial Browsers (i.e. in our case Google Earth) where they can generate simple 3D building geometries based on 3 parameters:

1. The footprint of the building (to be taken of the floor plane)
2. Number of stories of the building
3. Height of the stories

which will be provided by the user interaction (in run-time).The generation (of building geometries within 3D Geospatial Browsers) carries the potential of facilitating, mainly the joint decision making session (held physically or virtually) in evaluation of design proposals. In this case the interactive generation of building geometries will help in conducting a what-if analysis for evaluation of alternative designs to discover;

"How a design proposal would affect an urban scene"

and

"How the urban fabric would affect the detailed design of the building"

The digital city model containing geometrical representation of the current buildings in the scene (see the Section 3), will also act as a very important background data source for these analysis.

Methodology of the Implementation

In technical terms, there are two alternative routes for implementation, and they both share the same starting stage i.e. deciding on i) location for generating the geometry of the footprint, ii) number of stories and storey height. Then the building geometries can either be;

- Generated and visualised on-the-fly and stored later.
- Generated and stored in the first stage and visualised later.

These two alternative routes are illustrated in Figure 4.

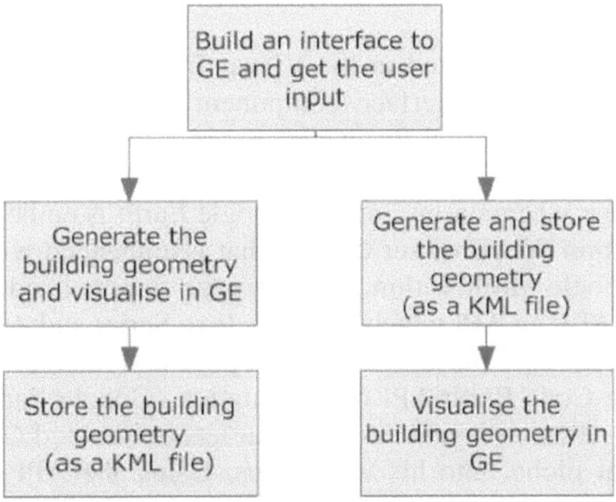

Fig. 4. Routes for generation of Building Models based on user interactions

The method chosen for implementation will not cause any technical difference, i.e. both implementations will provide the same output, in fact, generating the geometries on-the-fly and storing them in a later stage might bring performance efficiency for (the time behaviour of) the software.

In terms of municipality-citizens communication, the first option can be applied for interactive discussions, while the second one could be used by the citizens to prepare their ideas prior the discussions. In this primary development effort, the first method is chosen for the implementation. The implementation started with a prototyping effort to produce an early prototype for testing and validating the (technical dimension of the) implementation. The following section explains the details of this prototyping effort.

The Prototype

The prototyping effort includes three stages, the first stage was about the development of the (interaction) interface -for getting the user input-, the second stage involved the development of the component that generates the building geometry on the fly, and final stage is on development of the component for storing the generated building geometry. As the prototype is still in development at the time this chapter is prepared, the following sections will provide details of the first two stages that have completed until today.

The User Interface

A Windows Form, created in the Visual Basic.NET ® environment is used as the main user interface component of the prototype. A Web-Browser Control® of Visual Basic.NET is used, to embed a Web Browser within the user form. Then on the initialisation of the prototype application, a web page where an instance of Google Earth is embedded is called, and loaded within WebBrowser Control that is embedded within Windows Form. The Google Earth within the web page is populated by the use of Google Earth API, an API which provides Java Script and AJAX functions to populate the Google Earth within a web page and interact with it.

The Google Code Earth API documentation (Google Earth API, 2009) describes this API as a JavaScript API that lets one embed Google Earth, a true 3D digital globe, into his web pages. Using the API one can draw markers and lines, drape images over the terrain, add 3D models, or load KML files. The Google Earth API allows building sophisticated 3D map applications that are working within web pages.

Sample Applications using the Google Earth API are provided at the Google Code Playground (Google Code Playground, 2009) which is an instructive resource for the developers. The application uses the ‘Event Listeners’ provided by the API to get the coordinates of the points -that the user clicks- on the Google Earth application. Once the user clicks on the Google Earth window,

- the data is first handled by the Event Listener (of the Google Earth API)
- then it is transferred to a dynamic attribute of the web page (that the API is embedded in)
- finally when the Listener of the WebBrowser Control (in Windows Form) notices this change(in the dynamic attribute of the web page), the value of this dynamic attribute of the web page (i.e the coordinates of the point that the user clicked) is transferred into the Visual Basic.NET application.

The Windows Form provides the user command buttons to activate and deactivate the Event Listener of the Google Earth API (Figure 5).The captions of these buttons will be refined in the next version of the tool. The deactivation function is needed to stop the transfer of the coordinates by user clicks, when the user interaction for drawing the floor plan of the building is completed.

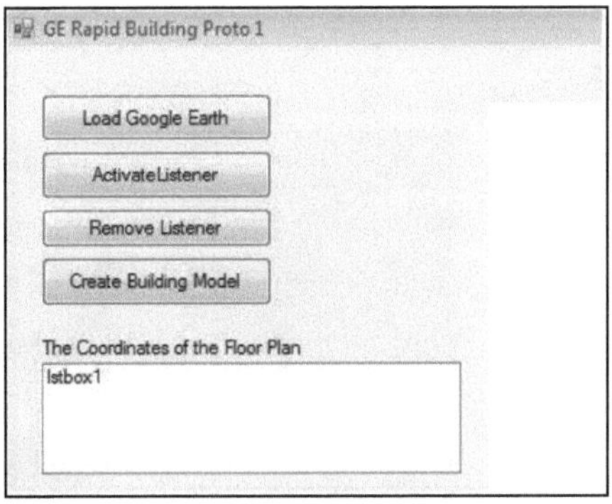

Fig. 5. The command buttons of the User Interface

These activation and deactivation functions are accomplished through Visual Basic.NET function calls to Java Script functions of the (of the Google Earth API).For example, the following Visual Basic.NET code line

```
wbrowser.Document.InvokeScript("addEventListen")
```

invokes the following Java Script function,

```
function addEventListen(){
google.earth.addEventListener(ge.getGlobe(),'click', getLatLonAlt);
eventListenerActive = true;}
```

in the web page, where the Google Earth is populated. By this way, the Google Earth API adds an Event Listener to the populated Google Earth instance. Following this, when the user clicks on the Google Earth instance the following function handles the event and transfers the values of the coordinates(of the location that the users clicks) into a dynamic attribute of the web page with the function below.

```
function getLatLonAlt(event) {
lat= event.getLatitude();
document.getElementById("mylat").value=lat;}
```

These, i) call of functions from Visual Basic.NET to Java Script and ii)passing values from the Java Script to the Visual Basic.NET application made it possible to acquire the user input (i.e. the coordinates of the floor plan) by user interaction in order to generate the building geometry.

The state of the User Interface after the user input is acquired is depicted in Figure 6. The List Box in the Windows Form now shows the coordinates of the (points that depict the) floor plan which are acquired by the user interaction with the Google Earth instance.

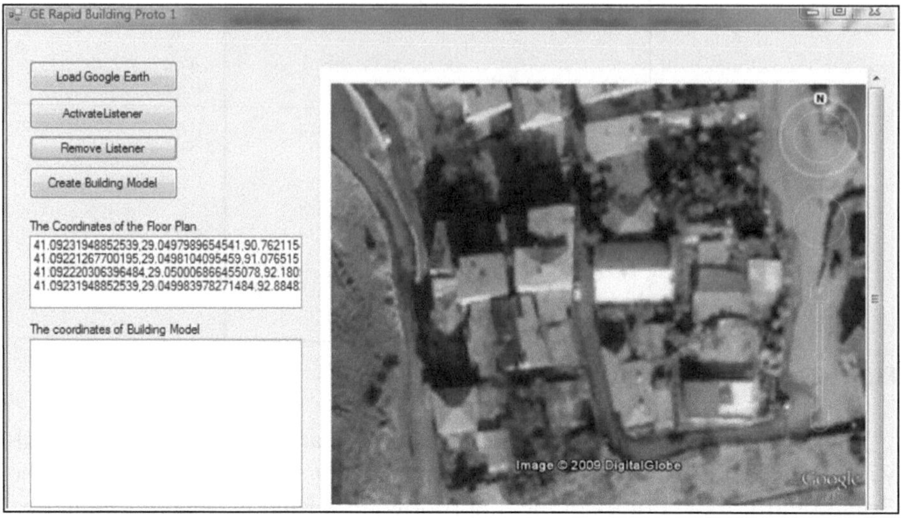

Fig. 6. The state of the User Interface (i.e. showing the coordinates of the points determined by the user interaction).

Generating the buildings

The building model generation functions are triggered by the "Create Building Model" command in the Windows Form (Figure 6). In the first stage (i.e. when the button is clicked) the user is asked to input the number of stories in the building and the storey height.

This information is acquired from the user and these values are passed onto the Visual Basic.NET functions that are used to generate the 'sides' of the building. In this first version, of the prototype we have made an assumption (on a conceptual simplification) that the building is represented as a rectangular-shaped one, where the floor plan consists of four points and every storey has four-sides (see Figure 7).

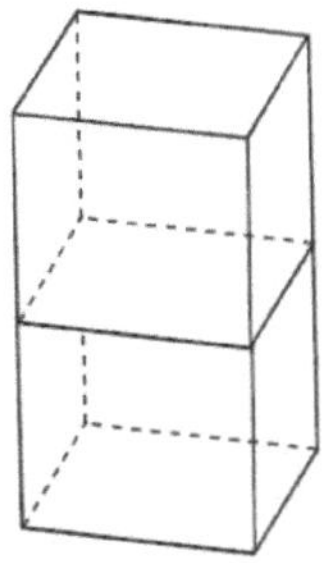

Fig.7. Illustration of Building Representation

The sides of the building are generated by Visual Basic.NET functions that call the Java Script functions of the Google Earth API. For example, the code for the Visual Basic.NET function that draws a side of each storey is provided in the following.

```
Function DrawSide()
wbrowser.Document.InvokeScript("polygonDraw", New Object() {False, False})
wbrowser.Document.InvokeScript("outerDraw")
wbrowser.Document.InvokeScript("outerAddPoint", New Object() {vvlat(1), vvlon(1), vvalt(1)})
wbrowser.Document.InvokeScript("outerAddPoint", New Object() {vvlat(2), vvlon(2), vvalt(2)})
wbrowser.Document.InvokeScript("outerAddPoint", New Object() {vvlat(6), vvlon(6), vvalt(6)})
wbrowser.Document.InvokeScript("outerAddPoint", New Object() {vvlat(5), vvlon(5), vvalt(5)})
wbrowser.Document.InvokeScript("outerFinalise")
wbrowser.Document.InvokeScript("polygonFinish")
End Function
```

The algorithm uses the *Multi Polygon Representation* approach when generating the geometry of the building elements. The side of each storey is represented with a polygon, which can contain any number of inner and outer rings. In order to generate the polygons, standard Java Script functions provided by Google Earth API are customised (to be invoked within the different Visual Basic.NET functions of the application). The code below demonstrates some of the Java Script functions that are called within Visual Basic.NET functions when generating the polygons.

```
function outerDraw() {
outer = ge.createLinearRing('');}
function outerAddPoint(lat,lon,z ){
outer.getCoordinates().pushLatLngAlt(lat,lon,z);}
function outerFinalise(){
polygon.setOuterBoundary(outer);}
```

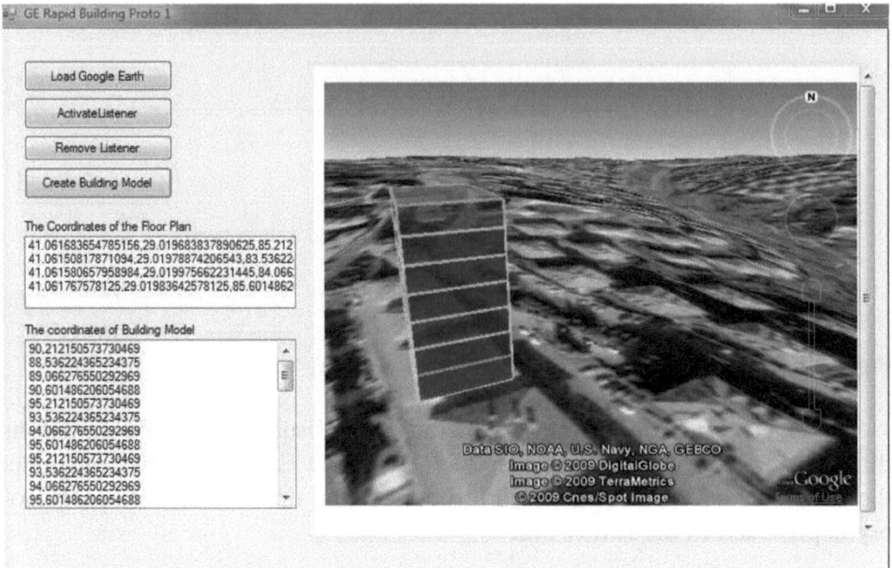

Fig. 8. A building model generated as a result of user interaction.

In this early prototype, the geometry of the generated buildings is very simple (i.e. every storey of the building is represented with a rectangular prism), but the implementation (even in this early form) provides the technical proof-of-concept that it is possible to rapidly generate building geometries within Geospatial Browsers based on the user input. Figure 8 shows an example of building models generated by the prototype application.

Conclusion

This paper presented a rapid visualisation tool for supporting the joint decision making in evaluation of design/renewal proposals. As discussed, the process involve participation of the citizens and experts. These parties have mostly different point of view and conflicting expectations from a development. The visualisation of the geospatial environment, the urban fabric and the proposed building/facility is the key to facilitating the joint decision making during the design evaluation process. The prototype presented in this chapter also technically proves the idea of using 3D Geospatial Browsers for enabling a user interaction. The developed prototype demonstrates that the simple 3D models of the buildings can be generated by user interaction with Google Earth. These (generated) models can then be used in evaluating design proposals (i.e. to determine how the proposed building will affect the urban fabric and also to determine how the surrounding environment will shape the final design of the building.). Planners and citizens can jointly examine the proposed designs or suggest their own in a physical or a virtual meeting. In this respect, the tool can be seen as a candidate to replace the simple 3D physical (wooden, plastic) models. As the tool enables the generation of simple 3D geometries only by using a well known Geospatial Browser it eliminates the need for learning to use a third part software (such as SketchUp or 3D Studio Max) to produce building models, which would require a reasonable time and effort especially for the citizens.

The prototyping effort will continue with the development of the component that will allow for generating more complicated geometries and storing these geometries as KML files. The prototype will then be tested with users in beta stage, of the development and feedback obtained from the validation studies will be used in developing the next version of the prototype. Another important step will be the integration of 3D models (of the existing buildings) with the (proposed) design models. Then the users will be able to create the new buildings directly in the environment containing 3D models of existing buildings.

References

Al-Kodmany, K. 2002, Visualization Tools and Methods in Community Planning: From Freehand Sketches to Virtual Reality, *Journal of Planning Literature* 17(2):189-211

Arayici, Y. 2007 An approach for real world data modelling with the 3D terrestrial laser scanner for built environment, *Automation in Construction* 16 (6): 816-829

CommunityViz 2009. The Web Site of CommunityViz Public Participation Software (available at http://www.placeways.com)

ESC. 2009 The Web Site of Environmental Simulation Center (available at http://www.simcenter.org)

Google Code. 2009 KML Tutorial, (available at http://code.google.com/intl/tr-TR/apis/kml/documentation/kml_tut.html)

Google Earth API. 2009 Google Earth API Documentation in Google Code, (available at http://code.google.com/intl/tr-TR/apis/earth)

Google Code Playground. 2009 (available at http://code.google.com/apis/ajax/playground/?exp=earth)

Hudson-Smith, A., Evans,S., Batty,M. and S.Batty, 2002, Casa Working Paper 60, Online Participation: The Woodberry Down Experiment, http://www.casa.ucl.ac.uk/publications/workingPaperDetail.asp?ID=60

Isikdag, U. 2006. Towards the Implementation of Building Information Models in Geospatial Context, PhD Thesis, University of Salford, UK.

Kang, Z., Z. Zhang, J. Zhang and S. Zlatanova, 2007, Rapidly realizing 3D visualisation for urban street based on multi-source data integration, in: Li, Zlatanova&Fabbri (Eds.) *Geomatics Solutions for Disaster Management*, Lecture Notes in Geoinformation and Cartography, Springer-Verlag Berlin, Heidelberg, pp. 149-163

Kibria, M.S. 2008, Functionalities of geo-virtual environments to visualize urban projects, Master's Thesis GIMA (Utrecht Univ., TU Delft, Wageningen Univ., ITC), 2008, 124 p (available at www.gdmc.nl/publications).

Kibria, M.S., S. Zlatanova, L. Itard and M. van Dorst, 2009, GeoVEs as tools to communicate in Urban Projects: requirements for functionality and visualization, In: J. Lee and S. Zlatanova (Eds.); 3D Geo-Information Sciences, Springer, 2009, pp. 379-395

Knapp, S. and V. Coors, 2008, The use of eParticipation systems in public participation: The Veps example,in: Coors, Rumor, Fendel&Zlatanova (eds.) Urban and Regional Data Management, UDMS Annual 2007,Taylor & Francis, Leiden, The Netherlands, pp. 93-104

Kolar, J., T. Bayer and S. Grill, 2008, Open source GRIFINOR platform for 3D spatial data presentation. In: Remote Sensing for a Changing Europe: Proceedings of the 28th Symposium of the European Association of Remote Sensing Laboratories, Istanbul, Turkey, 2-5 June 2008

Pu, S. 2007. Automatic Building modelling from terrestrial laser scanning, In P.Van Oosterom, S. Zlatanova, F.Penninga, E. Fendel (eds). *Advances in 3D Geoinformation Systems*, LNG&C,Springer, pp.147-160.

Reitz,T.,Kramer,M., and S.Thum, 2009, A Processing Pipeline for X3D Earth-based Spatial Data View Services, In Proceedings of the 14th International Conference on 3D Web Technology ,ACM, pp.137-145

Tao, V. 2006. 3D Data Acquisition and object reconstruction for AEC/CAD, in Zlatanova & Prosperi (eds.) *Large-scale 3D data integration- Challenges and Opportunities*, Taylor & Francis Group, CRCpress, Boca Raton, pp.39-56

Zlatanova, S., L. Itard and M. Van Dorst, 2008, User requirements for virtual environments used to model buildings at the urban scale, in proceedings of IBSA-NVL 2008 Event, 9 October, 2008 (CD)

An Experimentation of Expert Systems Applied to 3D Geological Models Construction

Eric Janssens-Coron[1], Jacynthe Pouliot[1], Bernard Moulin[2], Alfonso Rivera[3]

[1] Geomatics Department, Pavillon Louis-Jacques Casault, Université Laval. G1K 7P4, Quebec, QC, Canada. eric.janssens-coron.1@ulaval.ca
[2] Computer Sciences and Software Engineering Department. Pavillon Adrien-Pouliot, Université Laval. G1K 7P4, Quebec, QC, Canada.
[3] Geological Survey of Canada. 490, rue de la Couronne. G1K 9A9, Quebec, QC, Canada.

Abstract. To address the challenge of sustainable exploration and exploitation of oil, gas, mineral or groundwater resources, engineers and scientists often exploit 3D geological models to visualize, assess and understand the complexity of underground systems. But the construction of 3D geomodels is time-consuming and it generally requires the use of specialized software and specific expertise, in particular in geology where the system is not fully visible and necessitates complex efforts of human conceptualization. In order to simplify this process we propose to assess the ability of an expert system to analyze a particular source of geological information - geological cross-sections - and to enable several technical steps to build 3D geological models. This paper presents the procedure used to enable the development of this new expert system applied to the context of 3D geological modeling. We will try to answer questions such as the kind of information that is required, the way to build a knowledge base, the limits of using such a system, the quality of the 3D model output, etc. The discussion is based on an experimentation done in collaboration with the Geological Survey of Canada by comparing the current 3D geomodel and the one produced by the prototype 3D GeoExpert.

Keywords: 3D modeling, geological modeling, expert system, rule-based reasoning

T. Neutens, P. De Maeyer (eds.), *Developments in 3D Geo-Information Sciences*,
Lecture Notes in Geoinformation and Cartography, DOI 10.1007/978-3-642-04791-6_5,

1. Introduction

When the sustainable development becomes an economic and social challenge, the exploration and the exploitation of underground natural resources have to develop highly specialized techniques of extraction and tools of analysis. Since fifteen years, new 3D modeling methods and software provide the possibility to geoscientists, experts in Earth sciences such as geologists and hydrogeologists, to build 3D geological models [1]. These geomodels can be used for improving the capacity for analyzing the underground systems for geoscientists, for facilitating the decision-making process for the management of these systems, or to enrich the experiment of visualization and the broadcasting of the information to the public.

Geologists are usually, but not always, the professionals who build 3D geological models. They perform specific and multifaceted analyses of geological data thanks to their geological knowledge. Despite their expertise, the geological modeling process is long and complex. The construction of a worthy 3D model can take several weeks or months depending on the size of the studied area and the complexity of the geological structures on the underground. Their knowledge of 3D modeling techniques and of specialized computer-aided design software (CAD) such as gOcad[1] has also an impact on the time spent for the construction and the reliability of the content of the model.

In order to help people to effectively and quickly build 3D geological information, we propose to couple an expert system and 3D CAD tools to produce automatic 3D geological representations. Because geological analysis is based on several principles and rules that can be converted into expert system rules, we chose to use a rule-based reasoning (RBR) system to analyze cross-sections and to build a 3D model. Section 2 provides a few explanations about the process of 3D geological modeling from cross-sections analysis. Section 3 presents a short overview of expert-systems and their applications relevant to geological applications. Sections 4 and 5 describe the 3D GeoExpert tool developed and its experimentation to 3D geological modeling of the study site of Saint-Mathieu's esker, Quebec (Canada). In section 6, we discuss the feasibility and interest of using an expert system for 3D geological modeling.

3D GeoExpert is specifically dedicated to the 3D geological modeling applied to hydrogeological applications. This research was part of the

[1] GeOlogical CAD with key component features including Log display, cross-section and map views and structural framework builder, see http://www.pdgm.com/gocad-base-module/products.aspx

GeoTopo3D Project[2] for the "development of a 3D Predictive Modeling Platform for Exploration Assessment and Efficient Management of Mineral, Petroleum and Groundwater Resources". The GeoTopo3D project was funded by the Networks of Centres of Excellence GEOIDE[3] (GEOmatics for Informed DEcisions).

2. 3D geological modeling and cross-sections interpretation

2.1. Introduction

As stated by Bédard 2006 [2], Massé 2003 [17], and Pouliot et al., 2003 [19], the methodology for the construction of 3D geological models can be divided into three generic phases: (1) Data setup and cleaning of objects in a 2D universe, (2) Insertion of the previous objects in a 3D universe, (3) Construction of continuous volumetric objects. Because direct measurements of 3D objects boundaries are difficult to carry out in the case of subsurface systems, data sources are often limited to geological maps, drillings and cross-sections to illustrate and understand the subsurface. Geological maps are 2D representations of geological formations outcropping on the ground surface, while drillings are wells drilled to determine the structure and the constituent materials of the underground.

A geological cross-section is a two-dimensional model highlighting, along the vertical axe, the sequence of geological formations with their characteristics as they are found either on the ground (geological area) or in polls (subsurface geology). It is composed of two types of geological objects: units and contacts. A geological unit is a set of terrains that can be individualized for tectonic and/or stratigraphic reasons. A geological contact is a generic term identifying boundaries between two geological units. It has a specific semantic value according to the geological event or process that has put it in place [10].

Any geological object is associated with spatial and geological information. Spatial information relates to the geometry of objects and the topo-

[2] http://geotopo3d.scg.ulaval.ca/
[3] http://www.geoide.ulaval.ca/

logical relationships between them. In cross-sections, geological units are represented by surfaces and geological contacts by curves. Geological information describes the geological nature of the object. For example, as shown in Figure 1, a surface is associated with a rock type in order to represent a geological unit while a curve is a contact (e.g. topography).

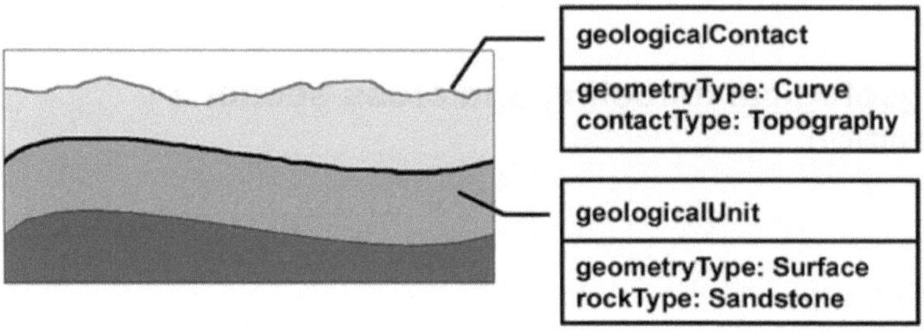

Fig. 1. Simplified geological cross-section with three geological contacts and three units. Each object has its own spatial (geometryType) and geological (rockType, contactType) attributes.

Some regions that have undergone deformation may have complex geological structures. In these cases, contacts may be more complex because they can define sets of geological units. They have a particular terminology such as faults, for example.

2.2. Spatial information and spatial data model

To identify precisely the information contained in geological cross-sections, we have conducted several interviews with geologists from the Geological Survey of Canada (GSC) in Quebec and with non-experts. One of the objectives of these interviews was to evaluate what elements (geometry, topology, graphic semiology, etc.) are used during the reading of a cross-section and what kind of information the reader extract. This study has highlighted the importance of geometry and especially topological relations between objects in the understanding and construction of cross-sections. This observation is also true in the case of 3D models. The proposed data model (Fig. 2) is based largely on the spatial schema of ISO-19107 [13]. The spatial data model can be used to represent the spatial information included in geological objects constituting drillings (1D), cross-sections (2D) or the whole 3D model. In the context of cross-section analysis we are mainly interested in properties such as shape, connectedness and orientation.

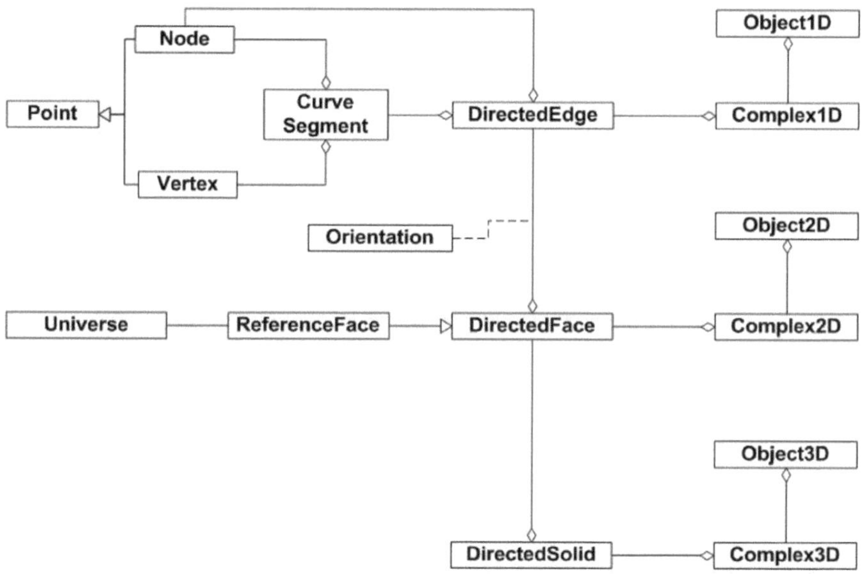

Fig. 2. The spatial data model used for the representation of objects of 2D geological cross-sections and 3D models

In this spatial data model, the 0D geometric primitive is the point (Point) and corresponds to a triplet of coordinates (X, Y, Z). The Node and Vertex are the 0D topologic primitives (TP). Vertices are used only when a contact is represented by a set of straight segments and thus connect two segments of a same contact. Nodes are extremities of a contact, whatever the modeling type used (straight segments, Beziers or Nurbs curves). DirectedEdges are the spatial component of simple contact while complex contacts such as faults are represented by Complex1D (Fig. 3). The association of the spatial component and geological properties (e.g. the type of rock) is identified by the feature class Object (1D to 3D).

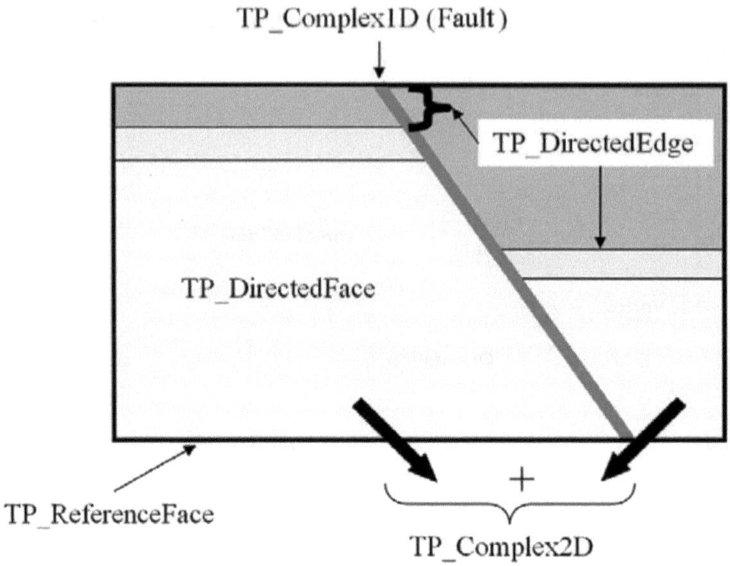

Fig. 3. Examples of possible encoded elements in the proposed spatial data model

While DirectedEdges are oriented features in ISO-19107, DirectedFaces can be defined by the set of DirectedEdges that defines their border. DirectedFaces are also oriented. ReferenceFace can identify DirectedFaces in a universe that may be 2D, the cross-section, or 3D. Complex2D and Complex3D are the spatial components of Object2D and Object3D that are sets of geological faces or solids having the same geological properties.

It is important to note that this choice of spatial primitives leads to a discrete representation of the 2D (cross-sections) and 3D (3D models) space and not a continuous representation in which the faces and volumes would be filled and made of pixels (2D) or voxels / tetrahedra (3D). This kind of geometric modeling strategy based on boundary representation (B-Rep) was considered optimal, taking into account the subsequent integration of the geological model with numerical modeling methods (e.g. finite element) widely used in hydrogeological applications.

2.3. Geological information

Because geological cross-sections use both geological and spatial information, we selected GeoSciML, a domain ontology for geoscience information [21]. The GeoSciML project was initiated in 2003, under the auspices of the CGI, the Commission for the Management and Application of Geoscience Information, a commission of the International Union of Geological Sciences. It establishes a common set of feature types based on geological criteria (units, structures, fossils) or artifacts of geological investigations (specimens, sections, measurements). GeoSciML was initially defined to interpret geological maps, but it can be extended to geological cross-sections since they are maps of vertical sections of the underground. We didn't use the whole GeoSciML ontology. We restricted its use to the lithology which is a critical geological criterion in the analysis of cross-sections.

2.4. Building geological cross-sections

Cross-sections are the result of the interpretation and integration of several types of data (from the morphology of the landscape to geophysical data) done by geologists [8]. It explains the huge volume of information contained in cross-sections.

To build cross-sections, geologists use a wide range of knowledge in several domains: stratigraphy, structural geology, petrology, etc. to analyze these data. Some of that knowledge is expressed in the form of rules. Among these rules, the principles of stratigraphy defined by Steensen in the 17th century are the basic for the construction of geological cross-sections and even 3D digital models. They are defined as follows [4]:

- *The principle of superposition* states that "strata become younger upwards – unless the entire section has been overturned during deformation".
- *The principle of initial horizontality* states that "strata were initially deposited flat".
- *The principle of original continuity* states that "rock layers were originally continuous. For example, horizontal strata exposed in river valleys were once connected and have since been eroded by the river cutting them".
- *The principle of cross-cutting relationships* states that "a tabular layer is younger than any layer it cuts across".

These principles allow geologists to understand the temporal sequences of the geological events that have driven to the geological structures observed today. They are essential to build geologically and topologically consistent models but also to analyze cross-sections and extract useful information. That's why such principles are an essential part of the knowledge used by the expert system.

3. Geological cross-sections analysis by an expert system

3.1. Rule-based expert system

An expert system is software that uses knowledge to simulate the reasoning of a human expert. It consists of a knowledge base, an algorithm for interpretation and use of knowledge and an interface to dialogue with the user [12].

The tricky part in building an expert system is the identification, acquisition and formalization of knowledge stored in the knowledge base [15]. There are two reasons for this. The first one is that the analysis of geological cross-sections and construction of 3D geological models require a very huge volume of knowledge in several domains: geology, geomatics (2D/3D spatial analysis, 3D modeling), and 3D geological modeling. We chose to restrain our work to the lithology and geological structures of Quaternary glacial deposits. The second difficulty in the knowledge acquisition process is that an important part of the knowledge used by experts in their reasoning is tacit [9]. It is necessary to make several iterations to identify these unwritten concepts and rules [18].

Several previous work in geology [6, 16], geological cross-sections analysis [7] and 3D modeling [1, 3, 20] have shown that a large part of geological knowledge can be expressed as rules or principles in the form of a conditional statement such as "IF ... THEN ...". For example, the principle of superposition can be converted in "IF the geological unit A is below the geological unit B THEN A is older than B". We extended this work by integrating geological knowledge used in 2D and 3D in a first generation expert system using rule-based reasoning. In these systems, the knowledge base includes facts (e.g. 'The type of rock of X is sandstone') and rules. Rules use known facts in order to build new facts (e.g. 'IF a material is sandstone THEN the material is a rock of an aquifer'). These rules are triggered and linked by an inference engine to form new facts that are stored in the knowledge base and can be displayed via the interface to the user.

3.2. Method of analysis of a cross-section

A cross-section represents, at a given time, the result of the geological evolution of the underground through geological ages. Therefore, in order to build a model (cross-section or 3D model), geologists interpret the "geology" to be represented by establishing, between the different surfaces, total or partial order relationships based on chronology [3]. Then the objective of the analysis of 2D cross-section is to obtain a construction scheme including the chronological steps for the construction of the geological objects of the cross-section. It will then be possible to construct a similar scheme for the construction of the 3D model from the schemes of two geological cross-sections.

To build these schemes, we used several rules and principles and in particular the principles of superposition and of initial horizontality. We converted them into rules and added them into the knowledge base among other rules we developed and that are used for the identification of spatial and geological facts. Finally, about sixty rules have been written. The expert system begins the analysis process with the automatic identification of the spatial objects: Nodes, DirectedEdges and DirectedFaces (Fig. 4).

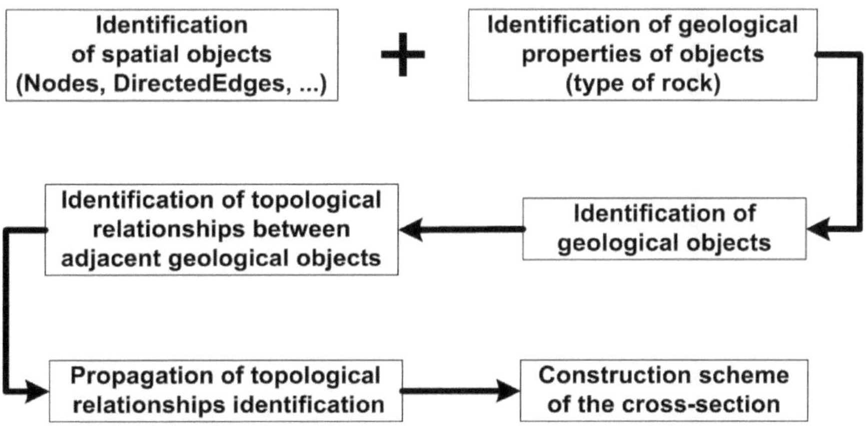

Fig. 4. Steps in the process of analysis of a geological cross-section

The important geological properties (e.g. the type of rock) are also identified, with the help of the user, via the graphic user interface of the system. The system checks that the terms used by the user are in accordance with the GeoSciML vocabulary stored in its knowledge base. Geological properties are then associated to the spatial objects in order to constitute geological objects. The next step is the identification of simple topological relationships between adjacent geological units and contacts (e.g. 'the unit U is adjacent to contact C'). Since the objects are referenced and oriented in the universe of the cross-section, it is possible to establish topological relationships such as "The unit U is below the contact C". By spreading the adjacency relationships with neighboring objects and using the principle of superposition the expert system finally obtain the scheme for the construction of the cross-section (Fig. 5). This scheme is a directed graph where geological objects (contacts and units) are nodes linked by relations of type 'below'.

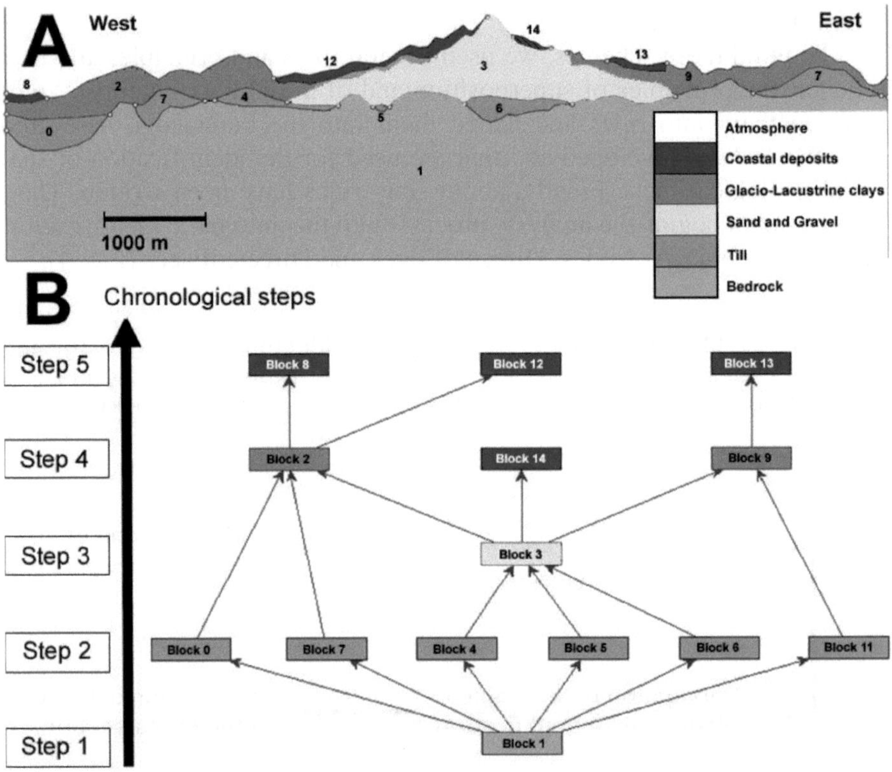

Fig. 5. A construction scheme of a cross-section. (A) A geological cross-section. (B) The resulting graph with the chronological steps of construction. To facilitate reading, only the geological units are represented.

For example, in Figure 5, the unit 3 (step 3) must be built before units 2, 9 and 14 because it is below them and it must be built after units 0, 4, 5, 6, 7 and 11 because it is above them.

4. 3D model building

The next step consists to build the 3D Geomodel from interpreted and integrated cross sections. The expert system uses two parallel geological cross-sections for which it knows the building schemes. The expert system builds the construction scheme of the 3D model by matching geological objects of the schemes in accordance with the chronological steps. The construction scheme of the 3D model is an oriented graph similar to the cross-sections schemes.

To build volumetric objects, the expert system needs to use an interpolation method. The choice of this method may have an impact on the construction of the building scheme of the 3D model [3]. We decided to choose a method using the top profiles of units that relies on the principle of original continuity. In a cross-section, a top profile is constructed by joining all the top contacts of the blocks of a same geological unit. A top profile exists along the entire length of the cross-section (Fig. 6A). A profile can be built only when all blocks of the unit exist. For example, in Figure 5, the profile of coastal deposits is not built in step 4, despite the existence of block 8 (the bloc in the center) because blocks 8, 12 and 13 are constructed only in the Step 5.

In the case of discontinuous units (i.e. units composed of several separate elements), or units that don't span the entire length of the cross-section (Fig. 6B), it is necessary to use a method to identify the contacts to be used to build the missing sections of the top profile of the unit. This method is based on the relative age of geological units. The top profile of a unit A cannot integrate geological units younger than the unit A, but only contacts of older geological units. If several contacts can be used to construct a missing section, then the contact with the geological unit immediately prior to the unit A is selected. For example, in Figure 6B, to build the top profile of the unit of Sand and Gravel, we use the contacts with the unit Till which is immediately prior to the sand unit (Fig. 6B to the west and east of the Sand unit), then contacts with the Bedrock unit in the sections where there is no Till (Fig. 6B to the east of the sand unit). Since the cross-sections are represented by graphs, the expert system can use the Dijkstra's algorithm to create the missing sections of the profiles of geological units.

Dijkstra's algorithm finds the shortest path in a weighted graph. In the case of the graph of a cross-section, the weighting depends on the relative position in the graph of contacts in relation to the top profile of the unit under construction. The contacts located above the unit (in a later chronological step) have an infinite weight. The contacts below the unit (in an earlier chronological step) have a weight proportional to their distance to the unit. For example in Figure 5, for the construction of the higher profile of the unit 3 (sand + gravel, step 3), the contacts built in steps 4 and 5 have an infinite weight. The contacts built in step 2 have a weight of 10 and contacts built in step 1 have a weight of 20. The end result is the profile of Figure 6B. The profile is above the sand unit (center of the cross-section) and passes over blocks of the Till unit and below the blocks of the Glacio-lacustrine clays unit.

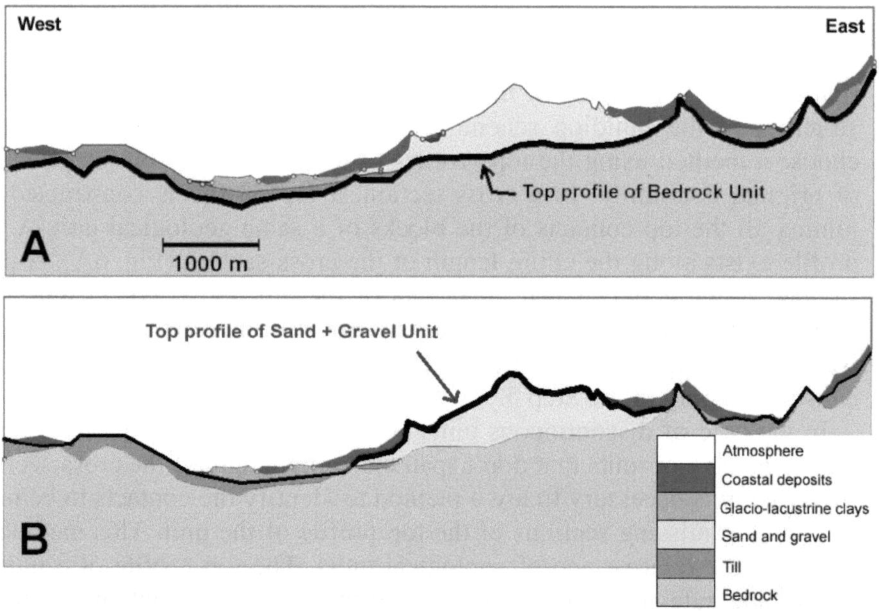

Fig. 6. Two top profiles. (A) Top profile of the Bedrock unit. (B) Top profile of Sand + Gravel unit.

Once the top profile of a unit is identified in each cross-section, the top surface of the 3D unit is built by joining two profiles by a B-spline surface. The top surfaces are built according to the chronological sequence of the building scheme of the 3D model. In some places, the top surface of the new unit penetrates into previous units although it is geologically and topologically inconsistent. The expert system has to clean these sections in order to obtain geologically consistent units. It uses topological operations such as intersection between surfaces and solids in order to carry out this task. The final 3D model is the set of geological contacts represented by surfaces and geological units represented by empty volumes limited by contacts (Fig. 7)

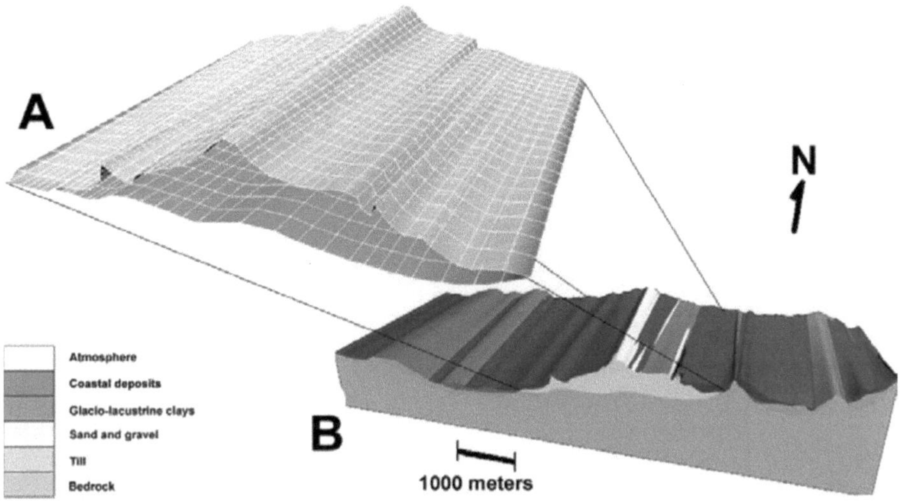

Fig. 7. A 3D geological model derived from our expert system. (A) The unit of sand and gravel. (B) The entire 3D model.

5. Experimentation

5.1. Site test

The experimentation was made on the site of Saint-Mathieu esker, in Quebec, Canada (Fig. 8). An esker is the geological result of melting glaciers. It is composed of moraine deposits and fluvioglacial forming a dam-like accumulation. The Saint-Mathieu esker has a length of about 120 km over a few kilometers wide. It is oriented roughly north-south and is located approximately 11km to the west of the municipality of Amos. It is the main source of drinking water of the city.

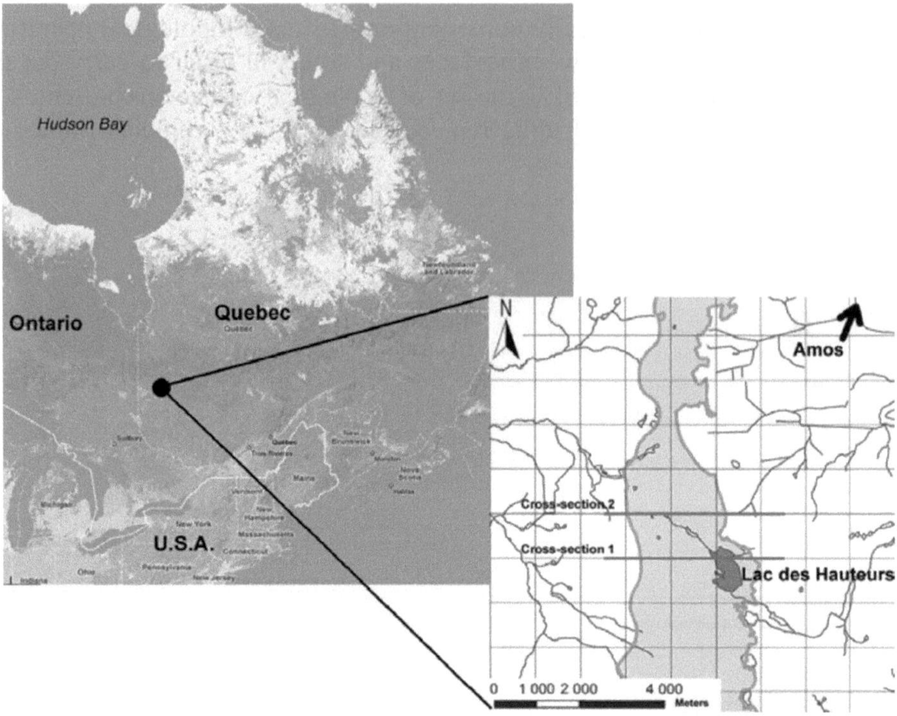

Fig. 8. Localization of the St-Mathieu esker, Quebec, Canada. The zoom shows the esker (in gray) and two cross-sections

We chose this site as a test site for three reasons. First, the geology is quite simple. There are no complex geological structures such as faults. This reduced the amount of knowledge to identify, particularly in geology. We limited our study of the Quaternary geology in a glacial context. Second, in the context of the GEOIDE project GeoTopo3D, we were interested to develop a 3D geological modeling tool facilitating the construction and the integration of 3D geologic models in hydrogeological models. Such an esker is representative of numerous local aquifers in Canada, which are established in porous media of glacial origin. Finally, the site was previously studied by the GSC. A geological model was built by the GSC experts using the 3D geological modeling software gOcad. The modeled area by the GSC is about 24 km long by 11 km wide [5]. This model was built from cross-sections and drillings. We used it as a baseline to assess the performance of our system and the quality of the output 3D geomodel.

5.2. Architecture of 3D GeoExpert

The prototype named 3D GeoExpert is composed of four elements: the expert system, a graphic user interface, a database and a 3D modeler (Fig. 9). It is mainly developed in Java.

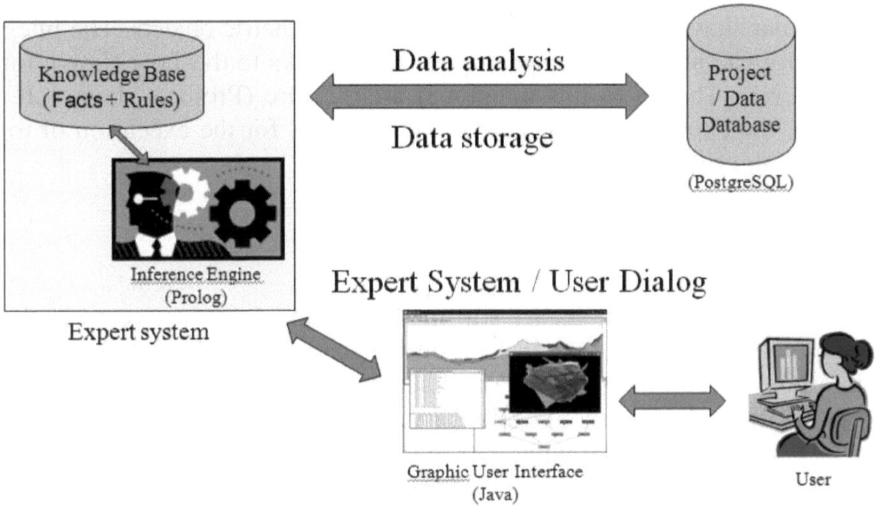

Fig. 9. Architecture of 3D GeoExpert

The rules and inference engine of the expert system was developed with Win-Prolog 4.7 (Logic Programming Associates ltd., www.lpa.co.uk). The graphic user interface (GUI) allows the expert system to dialog with the user in order to obtain missing data and to provide visualization capabilities. We used PostgreSQL v8.3 to manage the database that stored data: spatial and geological information from cross-sections and 3D models and general information about modeling projects such as the spatial reference system.

We used three external open source and free libraries. JGraph (www.jgraph.com) is a component extending the Java Swing library for easy graph visualization. We integrated it in the GUI for the visualization of the graphs of the cross-sections. JGraphT (jgrapht.sourceforge.net) is free Java graph library that provides mathematical graph-theory objects and algorithms. We used it especially for its implementation of the Dijkstra's algorithm. We used also the OpenCascade library for 3D visualization tools and algorithms of the topological operations needed for the construction of 3D objects. The OpenCascade library (OCC) is a software development platform developed in C++ and including components for 3D surface and solid modeling, visualization, data exchange and rapid application development. This huge library of several tens of thousands classes is freely distributed by OpenCascade S.A.S. (www.opencascade.org). Several geometrical and topological classes were used for the construction of the top surfaces of the units in 3D and topological operations (intersection, in particular) have been used to construct the volumetric objects. The integration of OCC into 3D GeoExpert was done thanks to the Java Native Interface (JNI). Thanks to this integrated architecture (Prolog – Java – JNI – OCC), the expert system is able to launch rules for the execution of topological algorithms.

5.3. Results

Since most cross-sections are still made on paper and then vectorized, 3D GeoExpert is able to exploit very simple spatial data. This data are stored in a text file as points (X,Y,Z) and segments linking two points. We used two geological cross-sections done by the GSC that are spaced about three km. This type of spatial input data allowed us to easily build geometric objects (Nodes, Vertices, DirectedEdges and DirectedFaces) according to the spatial data model. The acquisition of the geological information associated to the geometrical objects is done through the GUI. 3D GeoExpert asks simple questions to the user for each geological units and blocks such as what is the lithology or physical properties. After having collected all the necessary information, 3D GeoExpert begins the analysis of the two cross-sections and then builds the topologically coherent 3D model (Fig. 7 and Fig. 10).

The entire modeling process (cross-sections analysis and 3D geomodel building) takes about twenty minutes, depending on the amount of information to be provided by the user. The cross-sections analysis (about 450 Points, 60 Nodes, 90 DirectedEdges and 33 DirectedFaces) and the 3D geomodel construction (15 DirectedSolids) take less than a minute to process. In comparison, the manual construction of the same model from the same cross-sections that we performed in gOcad, took several hours of work.

The arrangement of units and contacts in our model is geologically correct although our model is visually different from the model of the GSC (Fig. 10). Two reasons explain these differences. On the one hand, up to now, 3D GeoExpert only exploits two geological cross-sections and a digital elevation model (DEM). The 3D geomodel produced by GSC was constrained by information coming from drillings data located between cross-sections. We are currently working in order to add drillings data to better constrain the shape of the objects. On the other hand, the interpolation method used in 3D GeoExpert differs from the one used by the GSC. The GSC built each block of each unit one by one while our method builds all the blocks of the same unit at once. This does not allow us to make straight forward comparisons on the number of objects into the models, for example.

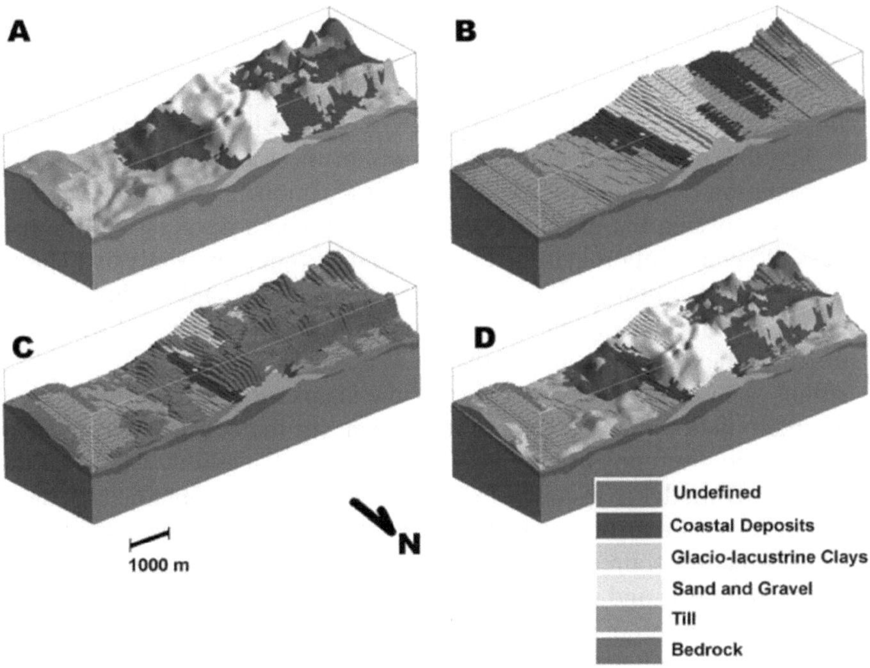

Fig. 10. Comparison between GSC's model and our model. (A) The GSC's model made in gOcad. (B) Our model obtained only with two cross-sections. (C) Our model obtained with two cross-sections and the DEM. The gray portion represents the terrain above the coastal deposits and under the DEM which could not be identified. (D) Superposition of the GSC's model (smooth surfaces) and our model (gridded surfaces).

6. Discussion and conclusion

The objective of this research project, conducted within the GEOIDE project, was to study the feasibility of an expert system to support the 3D geological modeling process. We have shown that it is possible to design an expert system using geological and spatial knowledge in a way similar to those used by human experts. 3D GeoExpert is able to analyze a certain type of geological data, geological cross-sections, and to build a 3D model from two geological cross-sections. As far as we know and based on our literature review and GSC networking, such a system does not exist. As a future work it would be interesting to couple 3D GeoExpert with a 3D geological modeling tool such as gOcad to assess their interoperability.

The use of directed graphs to represent the topological and geological information contained in geological cross-sections and to build 3D models allows applying the graph theory. Previous works have shown the usefulness of such an approach for geological cross-sections analysis [7] and for piloting and building 3D models [3, 20]. We extended this method to cross-sections analysis in order to prepare the construction of the 3D model. In particular, we showed that it is possible to use graph search algorithms such as the Dijkstra's algorithm. Thanks to this algorithm, we built an interpolation method easy to implement for the construction of volumetric objects from objects of two geological cross-sections. Although the resulting 3D model is visually different from the model obtained by the GSC with gOcad, it constitutes a solid base from which it is possible to add other information sources such as drillings. They would better constrain the shape of objects and allow obtaining more realistic geological objects.

The integration of an expert system in the 3D modeling process may offer several advantages including a reduction of the time of construction of 3D models. In addition, it can facilitate access to 3D geological modeling for users who do not have as extensive knowledge in geology than geologists. These are the two main contributions of our work. However, one of the main difficulties is the identification and acquisition of the knowledge needed for the geological analysis. For this reason, an important and distinguished part of our work was to develop the concept of modeling context. During our interviews with experts we found that such a context is very useful to identify the objectives of the model and the method that an expert, or system expert, need to apply to build the model. There are two sub-contexts in the modeling context: applicative context and geological context. The applicative context defines the objective of the 3D geological model: "A geological model of what?" The answer of this question determines the geometrical type of objects and the geological properties that the model should integrate. In our research, we focus on geological models used for hydrogeological applications (e.g. availability of groundwater in an aquifer, traceability of a contaminant in an aquifer). Volumes represented by their boundaries are well suited for these applications. In addition, the integration of physical properties (e.g. porosity) would facilitate the integration of the geological model in numerical models used by hydrogeologists. The geological context indicates to the expert system rules that should be used and rules that are not necessary. In our research, we worked on a glacial structure without structural deformations such as faults, so the expert system doesn't apply the rule for detecting faults. The concepts of modeling context and sub contexts are interesting and fundamental because they allow dividing the problem of knowledge acquisition into smaller areas of study in order to handle a relatively small volume of knowledge for each of them. This allows better organization of the knowledge base by separating the general knowledge (e.g. principles of stratigraphy) from knowledge related to specific contexts and geological applications. This modularity could also be useful to easily extend the knowledge base and the functions of the expert system.

A second problem concerns the knowledge representation. This paper focused on geological knowledge that can be formalized by facts and rules. This type of representation is well suited to the basic principles of geology, but some geological knowledge may require a case-based representation. This type of reasoning is based on the memory of cases previously encountered by the expert system to meet a new situation [14]. This applies, for example, for the identification of brittle structures (e.g. faults) or ductile structures (e.g. folds). Besides, the small size of our knowledge base (about 60 rules) allowed us to develop the expert system directly in Prolog. However, a more advanced knowledge representation, such as conceptual graphs, should be needed to build a larger knowledge base [22].

However these difficulties severely limit the availability of software capable of intervening in all geological contexts. According to our experience, it would be preferable to limit the use of such a system to very specific tasks, thus reducing the volume and complexity of knowledge to acquire.

Concerning the quality of the output 3D model, it can be assessed from two points of view. The first one depends on the quality and the type of input data used to build the 3D model. In this first version, we chose to use two parallel geological cross-sections to build a preliminary model. These cross-sections define the limits of the 3D model. In order to enhance the quality of our model we wish to add other data sources, in particular drillings. They would allow a better constraint on geometrical objects. The second way of evaluating quality is to assess the fitness for use of our model by hydrogeological applications. Our model includes volumetric objects represented by their boundary surfaces and several physical properties. That should facilitate the integration of the geological model into numerical models used by hydrogeologists. Integration tests have to be done but this part of the process of the quality assessment is still in progress with the help of the GSC.

Acknowledgements

We wish to thank the GEOIDE Network and the Faculté de Foresterie, Géographie et Géomatique (Laval University) for funding, Ms. Andrée Bolduc and MM. Yves Michaud, Serge Paradis from the Geological Survey of Canada in Quebec City who have graciously contributed their expertise and provided their data and Ms Karine Bédard for her help on the standardization of the 3D geological modeling process. We also thank the reviewers for their relevant comments that helped us to improve this article.

References

[1] Apel M (2005) A 3D geological information system framework. Geophysical Research Abstracts p. 6.
[2] Bédard K (2006) La construction de modèles géologiques 3D: de la standardisation au tutoriel. Master thesis, Université Laval
[3] Brandel S, Schneider S, Perrin M, Guiard N, Rainaud J-F, Lienhard P, Bertrand Y (2005) Automatic building of structured geological models. Journal of Computing and Information Science in Engineering 5:138-148
[4] Brookfield M.E (2004) Principles of stratigraphy. Blackwell Publishing
[5] Bolduc A, Paradis S J, Riverin M-N, Lefèbvre R, Michaud Y (2005) Development of a 3D geomodel for eskers: the case of the Saint-Mathieu-Berry esker, Abitibi, Québec, Canada. Geological Society of America Annual Meeting, Salt Lake City, Utah
[6] Campy M, Macaire J-J (2003) Géologie de la surface : érosion, transfert et stockage dans les environnements continentaux. 2ème édition. Editions Dunod
[7] Chiaruttini C, Roberto V, Buso M (1998) Spatial and temporal reasoning techniques in geological modeling. Physics and Chemistry of The Earth 23:261-266

[8] Duprat A, Golé F, Rocroi J. P (1973) Exemples d'application de traitements automatiques à l'interprétation de sondages électriques. Geophysical Prospecting 21:543-559
[9] Feigenbaum E.A (1977) The art of artificial intelligence: I. Themes and case studies of knowledge engineering. Technical Report: CS-TR-77-621, Stanford University, Stanford, CA, USA
[10] Foucault A, Raoult J-F (1988) Dictionnaire de géologie 3ème édition. Editions Masson
[11] Fournier J-C (2006) Théorie des graphes et applications : Avec exercices et problèmes. Hermes Science Publications
[12] Harmon P, King D (1985) Expert Systems: Artificial Intelligence in Business. New York, NY: John Wiley & Sons
[13] International Standard Organization (2003) 19107 - Geographic information - Spatial schema.
[14] Kolodner, J (1992) An introduction to Case-Based Reasoning. Artificial Intelligence Review 6:3-34.
[15] Liou Y I (1992) Knowledge acquisition: issues, techniques and methodology. ACM SIGMIS Database 23:59-64
[16] Mallet J-L (2002) Geomodeling. Oxford University Press

[17] Massé L (2003) Développement méthodologique pour la modélisation géologique avec exemple pour la région de Moose Mountain, Alberta. Master thesis, Université Laval
[18] McGraw K L, Harbison-Briggs K (1989) Knowledge acquisition: principles and guidelines. Prentice-Hall Inc
[19] Pouliot J, Lachance B, Brisebois A, Rabaud O, Kirkwood D (2003) 3D geological modeling: Are GIS or CAD appropriates? In: Proceedings of ISPRS Workshop, WG II/5, II/6, IV/1 and IV/2 Joint Workshop on "Spatial, Temporal and Multi-Dimensional Data Modelling and Analysis", Quebec, Canada
[20] Schneider S (2002) Pilotage Automatique de la Construction de Modèles Géologiques Surfaciques," Ph.D. thesis, Jean Monnet University and Ecole des Mines de St Etienne, St Etienne, France
[21] Sen M, Duffy T (2005) GeoSciML: development of a generic GeoScience Markup Language. Computers & Geosciences 31:1095-1103
[22] Sowa J F (1993) Conceptual Graphs. In Perez S K, Sarris A K, eds., 1995. Technical Report for IRDS Conceptual Schema, X3/TR-14:1995, American National Standards Institute, New York, NY

Data validation in 3D cadastre

Sudarshan Karki[13], Rod Thompson[12] and Kevin McDougall[3]

[1]Department of Environment and Resource Management, Queensland, Australia, [2]TU Delft, the Netherlands, [3]University of Southern Queensland, Australia

Abstract. In a cadastre, the 2D parcel is nowadays correctly considered to be a special case of the 3D parcel because the rights and restrictions extend beyond the surface itself. Storing, representing and manipulating a true 3D parcel however has not yet been satisfactorily achieved because of constraints in data modeling and software development. Significant research has been done to identify the best ways to represent a 3D solid, with rigorous mathematical testing on the respective merits of alternative approaches. Software companies have come up with their own ways of storing and validating 3D data, mostly as extensions of the 2D concepts. However, validation rules of one software may not be acceptable within another software's validation environment. The validation itself can be specified in great detail but sometimes this leads to redundant, repetitive or unnecessary processing. Because of the high volume of data a typical organization may be expected to handle, it is necessary for the rules to be streamlined and efficient. In this paper, validation is initially approached to answer questions such as: what is validation? why it is necessary to validate?, and how do we validate?. Limiting the scope to the 3D geometry or spatial representation of a 3D cadastre, the paper takes a novel approach in identifying the various aspects of validation of a 3D cadastral parcel and identifies the critical validation factors. It examines the validity within individual parcels and the relationship between adjoining or overlapping parcels in 2D or 3D. Although it is difficult to ensure completeness of rules, critical validation rules are examined for each identified factor.

1. Introduction

The cadastre deals with land and the rights associated to it. It can be considered to be a combination of geometrical and attribute data which is regu-

T. Neutens, P. De Maeyer (eds.), *Developments in 3D Geo-Information Sciences*,
Lecture Notes in Geoinformation and Cartography, DOI 10.1007/978-3-642-04791-6_6,

lated and governed by the country's constitution and other laws for defined purposes. The technical and judicial framework assist in defining the cadastral framework along with other drivers like the land market and present land administration needs. In recent years there has been an increasing trend towards the unbundling of property rights and a demand for more efficient usage of our traditional 2D cadastre through 3D developments. Accommodating and assisting the growth of 3D cadastre as an integral part of the land administration is being largely driven by technology. In many countries, the judicial framework supports a 3D cadastre, while in some others it is yet to be developed. The technical aspects of storing, retrieving, manipulating 3D cadastre is yet to be developed at par with 2D cadastre. Much of the problem lies in defining a data model for 3D cadastre, the interactions with the existing database and data capture methods, and the range of possible shapes and combinations of 3D objects in existence at present and those likely to be in the future. Additionally, support for the subdivision or consolidation of these 3D objects, validation rules for checking the data before, during and after entry into the system, and the optimal validation rules for entry of the data to the system must be considered.

This paper explores the similarities and dissimilarities in the 2D and 3D cadastral data context, focuses on differences between validation of a single object and validation of a set of neighboring objects or partitions of space in the 3D cadastral context, and then narrows down to the optimal validation rules necessary for a 3D cadastre keeping in mind that mixed representations of both 2D and 3D can co-exist in the same dataset or jurisdiction.

1.1. Structure of the paper:

This paper is structured in six main sections: Introduction, Representations based on existing methods, Validation extended from 2D to 3D, Validation in 3D cadastral context, Representations based on ISO19152 LADM, Discussions and further research.

Following a brief introduction to the paper, Section 2 deals with methods of constructing 3D objects from simpler objects such as the tetrahedron, the composite solid object, the extruded object, and their applicability to the 3D cadastre situation in relation to their validation requirements.

Section 3 is a short discussion on validation extending from 2D to the 3D domain. Questions such as what is validation? why do we validate? and when do we validate are discussed briefly to emphasize the need for a validation.

Section 4 of the paper focuses on a 3D cadastral validation situation based on the internal geometry of a 3D parcel, the relationships of 3D parcels to the base parcel, relationships to other parcels, unique geometrical situations that might exist, effects of further processing of 3D geometry and certain entry level validations.

Section 5 deals with the geometrical structures and validation requirements of the newly developed ISO19152 Land Administration Domain Model (LADM).

Section 6 concludes the paper with a Discussion summarizing the paper, with a sub-section on Further Research which proposes work that needs to be extended and studied in depth.

2. Representation of a 3D parcel based on existing methods

Although significant work has been done e.g. (Kazar *et al*) on the validation of generic 3D geospatial objects, cadastral 3D parcels have validation requirements which are more restrictive in some regards (e.g. some 3D parcels may be required to be within base 2D parcels), but may be less restrictive in other regards.

This section deals with the various existing ways of constructing 3D objects, and the validation requirements for them. While the rest of the paper focuses on boundary representations to define a parcel or 3D object, this section discusses other ways of constructing 3D objects, which may provide jurisdictions with options to progress to a 3D cadastre based on their individual circumstances like legal requirements, existing data structure and proposed usage etc.

2.1. Objects constructed using tetrahedrons

The tetrahedron method of generating 3D objects (see Fig 1), has been employed in 3D GIS and extensively studied by Penninga et al (2006) and Rahman and Pilouk (2007). The same principles can be applied to create 3D cadastre using tetrahedrons.

Validation must ensure:

1. Individual tetrahedrons are not degenerate (4 disjoint points)
2. Tetrahedrons used to construct an object do not overlap –

The nature of the decomposition ensures that the object is unambiguously defined by a set of tetrahedrons, so the main objective is met, but there are several additional requirements.

3. Composite external faces that are intended to be planar should be planar. (see discussions below)
4. There must be no gaps between tetrahedrons within the objects (this is achieved via a topological representation; e.g referring to nodes via their id and not to repeat the coordinate values)

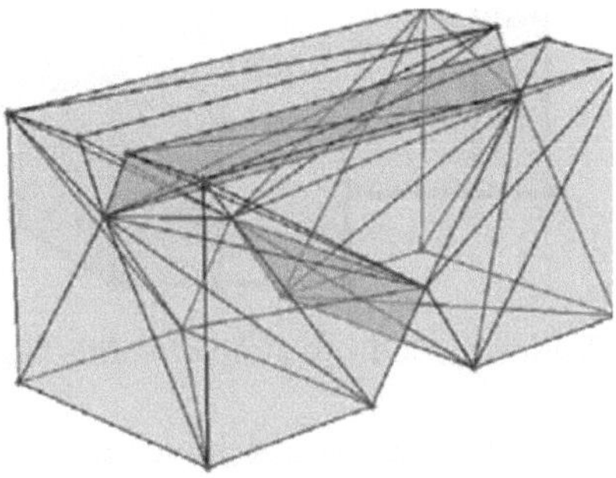

Fig. 1. A solid object created using tetrahedrons (Ledoux 2009)

***Discussions*:**

The validation rules to be considered after Penninga *et al* (2006) and Rahman and Pilouk (2007) are:

- where tetrahedrons are in contact, the contact may be at a single point only, along a line, or on a complete face. No partial contact should be allowed.
- where an object has planar faces, but these are broken into triangles by the tetrahedronization process, checks must be done to determine whether the original planarity is lost due to rounding errors of vertex coordinates.
- it may be considered important that a relationship is maintained between the faces to ensure that they can be kept coplanar (for example, the nearest faces of Fig. 1 clearly should be planar, but there is no indication to validate this).
- where vertices meet, the coordinates must be the same.
- there must be no duplicate vertices, which lead to degenerate tetrahedrons.

2.2. Objects constructed from simpler solids

The composite solid method of generating a 3D block (see Fig 2), can be employed for very low level representations of a 3D object or a 3D city model resembling LoD1 in CityGML (Kolbe 2009). A similar approach can be used to define 3D cadastral parcels.

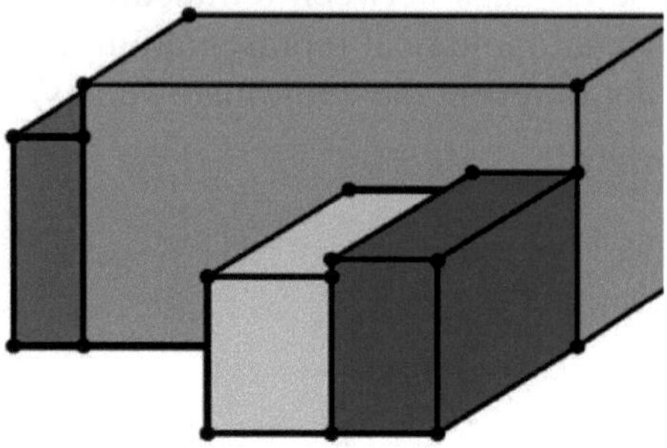

Fig. 2. Solids made from simpler solids (Ledoux 2009)

The simple solids produced would have the advantage of very simple validation constraints as well as quick answers to area and volume queries. It would also be easier to represent in a variable limiting height situation. Complex shapes can be generating by addition of simple volumes or subtraction to eat away from objects.

Where composite shaped volumes are built by eating away surfaces, making sure of the coplanarity. Gaps and intersections may be left during volumes addition or subtraction as well as in contiguous parcels, so that would have to be checked.

2.3. Regular Polytope

A particular example of the construction of objects from simple solids is the "Regular Polytope" (Thompson 2007, Thompson and Van Oosterom 2007). This approach uses as the simple solid a convex region of space, defined as the intersection of half spaces. A Regular Polytope is defined as the union of any number of convex regions (see Fig 2.1). The characteristic of this approach is that any combination of half spaces and convex regions is valid, greatly simplifying the validation process. All that is required is that the size and position of the objects are reasonable, and that continuity constraints are enforced if necessary.

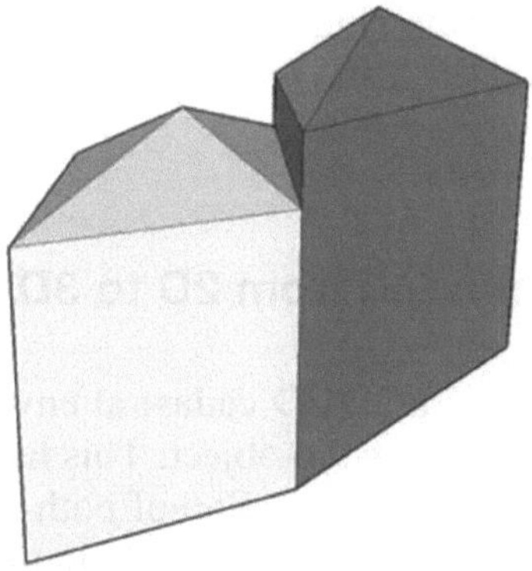

Fig. 2.1. Example of a regular polytope

2.4. Objects constructed by extruding

Three dimensional "objects" can be created using extrusion in many software. This is a convenient way to visualize 3D where variables such as individual object heights, constant heights for objects of same elevations and sometimes even the elevations of the land can be included to generate the 3D view. This may sometimes be the way to generate 3D cadastral views in non-complex situations, where no analytical facilities are required.

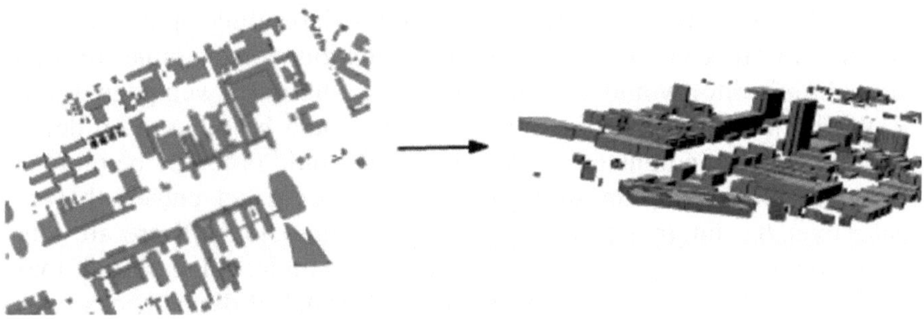

Fig. 3. City model using extrude (Ledoux 2009)

The extrusion method of generating 3D objects (see Fig 3), is fairly simple to build in a commercial software, but since the polyhedra formed are not topologically encoded, checks are limited to the 2D footprint and 3D data extraction is not possible except as a column of height attribute. This method provides a simple way of visualizing 3D cadastral data, but validation is limited to the geometry of the 2D base parcel and the height attribute of the extruded 3D object. However, if validation is required for adjoining objects of differing heights, it becomes more complex as demonstrated by Ledoux (2009). Also, it is more difficult to integrate 3D volumes with 3D terrain.

3. Validation, extended from 2D to 3D

The object of validation in a 2D/3D cadastral environment is to form a rigorous definition of what is a valid object. This has to be considered in 2D and 3D as well as a mixed representation of both in the cadastre. In the cadastre as well as in other applications, validation needs to be done at various stages of data entry and processing to ensure the integrity of the data and also that future analysis or operations are performed within their intended bounds.

Within the context of validation we need to answer the following:

- What is validation? (Concepts, approaches).
- Why we are validating? (Needs analysis, purpose of validation).
- How we are validating? (Validation rules).
- When should we validate? (At data capture, data entry, post processing etc).

Validation is the process of checking for possible errors in data via predefined rules usually before the data is processed or entered into the system. Validation takes on different meanings depending upon the context and circumstances of the application. Validation, for example, to a database administrator would be quite different to that for a web page developer. While the former's validation strategies would be focused on defining integrity checks and business rules, the latter would probably be more concerned about developing strategies to accept or reject entries that users make over the internet and may possibly be malicious. There are lots of common premises for validation as well, for example, data security, logical consistency, data within pre-defined range, format of data, checks for null values etc.

In digital cadastre, the need to validate arises from two simple questions: (i) who owns the particular land or space and (ii) what is the extent of what is owned. Thus, the major reason to validate is the need to provide unambiguous answers to these questions. Where 2D cadastre is in place the extent is generally defined, only limited by the scale of the data capture, the techniques used, the accuracies achieved etc, while the ownership information is tied to the parcel and stored by some means. The 3D cadastral situation is more complex as it gives rise to the primary problem of unambiguously defining the 3D parcel and its extent. Ownership information sometimes becomes complex with crossing networks, multiple strata etc.

Although there are similarities as well as dissimilarities in the approaches of the various validation requirements and strategies, the principles generally remain consistent. Generally, an attempt is made to identify all known issues then validation rules are defined against which data are checked before any further processing is undertaken.

This holds true in the 3D cadastral context as well. At this stage it is assumed that there is a 2D cadastre in place, which supports the key elements of storing and retrieving geometrical and attributes data in the system. Thus, storing of attribute information for a 3D parcel can be expected to raise few extra changes. It is the geometrical aspect that is highly complicated to store and validate before any manipulations can be performed.

Geometrical aspects are complicated in the 3D cadastral context because of the variety of shapes and geometries that are possible in a volumetric space. At present there is a general lack of consensus on how best to store or depict the data among the major software dealing with 3D data. For example, using the Oracle database one can define nodes with x, y, z coordinates and use them to construct edges, faces and solid objects.

As in the example in Fig 4, it would be possible to define a polygon from four coordinated points or nodes, and then a solid using the six polygon faces. However, for a truly vertical face, the vertical coordinates would be treated as duplicate points in some software's validation rules, which can disallow entry of the data into the database, thus creating interoperability issues.

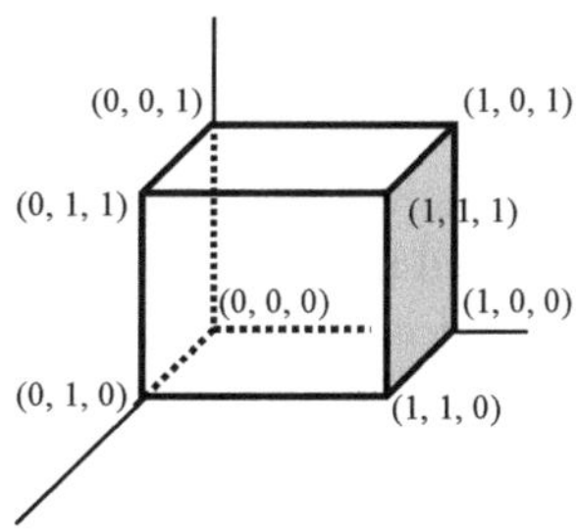

Fig. 4. Defining a volume by enclosing faces

The Open Geospatial Consortium (OGC) and the International Organization for Standardization (ISO) are working together to define a model, for generic 3D spatial information. The OGC (2004) and ISO19107 (2001) are examples of such standardizations.

Of particular interest in the field of cadastral data is the emerging standard ISO19152, the Land Administration Domain Model (LADM). This is an attempt to create an extensible standard that can be used to exchange cadastral information, and that can assist with the implementation of cadastral databases. (Lemmen et al 2009).

As an example of validation, ESRI's specification of ParcelFeatures_Topology (ESRI 2004) shows quite a comprehensive set of rules, tolerances, participating feature classes and the feature to be checked against. Other software have similar rigorous checking in the 2D GIS domain. This kind of checking is as yet to materialise in the 3D cadastre field because there is no agreed standard of validation of the 3D parcel and the ways of defining the rules are justified by the particular situation of the parcel itself (section 4).

Validation can be carried out in various stages during the life of the data. In the 2D or 3D cadastral context, validation can start from the time a surveyor sets foot on the site to the time it is entered to the system and might continue throughout the transaction life of a parcel. Validation strategies depend on the various stages the data goes through, and also in the case of a 3D cadastre the various methods or options of storing the data. In the following section, various options, methods and issues associated with a 3D cadastre are dealt with in further detail.

4. Validation in 3D cadastral context

In this section, situations involving 3D geometry in the cadastral context are listed. These are followed by discussions on the minimum validations rules or specifications that would be required to address these circumstances. In most cases, the 3D volume parcels are within the 2D base parcel, but exceptions may arise in the case of network objects when the 3D network object may extend beyond the base parcel.

The situations described below are classified into various groups based on similarity of conditions. They are based on:

1. Internal validity of 3D parcels.
2. Surface or base parcel.
3. Relationships to other parcels.
4. Unique geometrical situations.
5. Further processing on the geometry.
6. Entry level validations

4.1. Internal validity of 3D parcels

In the same way as it is necessary to ensure that 2D "polygons" have a valid definition, such as the simple feature specification by the Open Geospatial Consortium (OGC 1999), validity checks are necessary to ensure that representations of 3D regions are internally consistent. For example, in a boundary representation, the boundary must be complete, (watertight). Other issues like dealing with coplanarity, internal rings, constructing vertices, etc are equally important. These issues are explored in Kazar *et al.* (2008), to test for Closedness and Connectedness in particular.

4.2. Relationship to surface or base parcel

Unless the entire cadastre is to be represented in 3D, a mixture of 2D and 3D parcels will exist, with the vast majority being 2D. It is the nature of cadastral data that an identifiable set of parcels forms a base layer, which represents the primary interest in the land. In the following discussion, it is assumed that a set of base 2D parcels exists, and forms a non-overlapping partition of the area of the jurisdiction, and that appropriate validation and structural constraints are in place to ensure this. In many cases there will be secondary interests, like easements, represented by 2D parcels. These may form a second complete coverage but usually do not, and may be allowed to overlap.

In this section, the relationship between 3D parcels and the underlying base layer is explored, with reference to the validation tests that may be applicable.

Note that the term "base parcel" should not be taken to mean that it is physically below the "non-base" parcel(s). As noted by Stoter and van Oosterom (2006), the correct interpretation of what is known as a 2D parcel is in fact a column of space from some (possibly unspecified) depth below the earth surface to some (possibly unspecified) height above. What is usually represented as the 2D parcel is the intersection of this column with the surface. However, it must be noted that both bounded and unbounded objects exist in 3D space, for example, objects with open top or bottom.

Requirements

The relationship between the base parcel and the 3D parcels creates validation requirements such as:

1. Whether the 3D parcel is entirely within the base 2D parcel.

2. Checks on the upper and lower bounds of the 3D parcel (which may or may not have bounds on either ends).
3. Variable ground level in relation to the 3D bounds (where there are restrictions on the height above or below the surface to which rights or restrictions are imposed).

Discussions:

At least in Queensland, Australia, while the 3D parcels are initially created totally within the 2D parcels, (but this is not universal even in Queensland) this may not be preserved when the surface parcel is subdivided, e.g. a tunnel under a surface parcel does not prevent the owner of the surface parcel from subdividing it, as a result, requirement 1 above may not hold for all time.

The validation rules to make sure that the above conditions were met (if applicable) could include:

- *Ensure that the footprint of the 3D parcel is completely within the 2D parcel.*

  If the same horizontal stratum is considered, then it becomes a trivial 2D polygon check, however, as more 3D objects are compared to the base 2D parcel, the situation becomes increasingly complex as in the case of network objects.

- *Ensure that the extrusion above or below the surface by the 3D parcel was completely within the 2D parcel*

  If a building is considered to be built enveloping the total base parcel, then the mathematically vertical extrusion in given 3D reference system at right angles to the built surface might be different to the gravity vertical.

- *Ensure that the 3D parcel lies within the legal limits.*

  In some countries, there may be a restriction on the above or below ground rights of the 2D parcel. The 2D parcel may or may not be considered to be a special case of 3D as suggested by Stoter (2004), but would vary according to the cadastral interpretation in use for that particular jurisdiction. The LADM (ISO19152 2009) makes this interpretation explicit by defining the boundary of a 2D parcel as a "FaceString", which represents a set of vertical faces, but is actually stored as a 2D linestring (Fig 5)

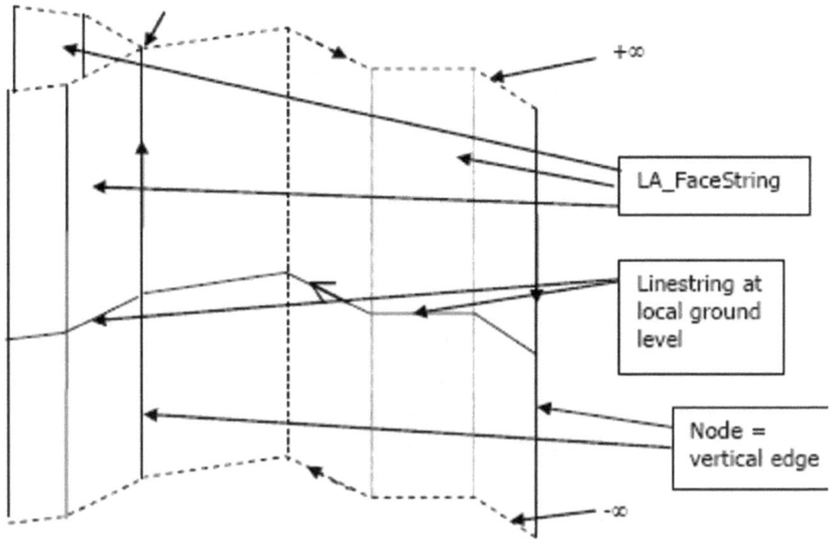

Fig. 5. Face string concepts of a 3D parcel based on 2D Line string at ground level ISO19152 (2009)

- *Ensure that there are no gaps or overlaps within the individual 3D units that are part of the 2D base parcel*

If the base 2D parcel contains several 3D parcels, as may be the case with buildings on a parcel, then it is just a simple check to see whether there are overlaps or gaps when they are supposed to be touching. However, in the case where network objects are treated as real estate and have defined extents, and surface parcels have defined depths, there might exist a gap between the two strata. This situation may be acceptable from an administrative point of view, but raises the issue of whether the remaining parcel, with the defined 3D parcels excised, is a valid construct (see Fig 6).

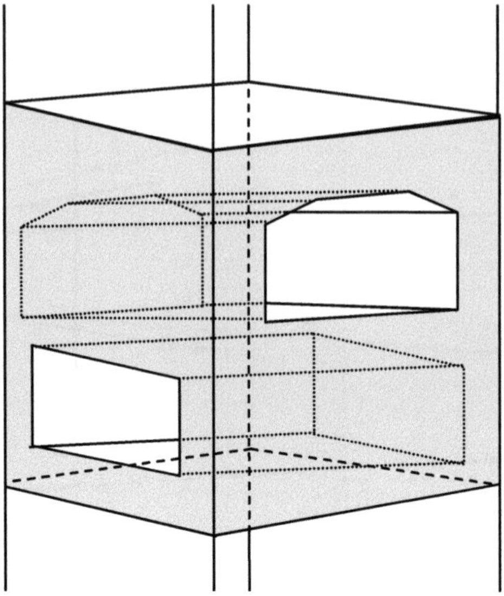

Fig. 6. Complex shape formed when two network parcels cross below, and are excised from a 2D parcel

4.3. Relationships to other parcels

Requirements

Beyond the relationship between 3D parcels and their enclosing base parcels, there is a need to consider the 3D parcels in relation to other parcels, including:

1. Checks on situations where a 3D parcel meets 2D parcels.
2. Checks on the geometry and topology of the 3D parcel and its relationships with neighbouring 3D parcels.

Discussions:

- *Ensure that there are no gaps or overlaps among neighbouring 2D or 3D parcels.*

When considering a simple 2D geometrical network of parcels, the obvious checks would be to find out if there were any (i) self intersections, (ii) overlaps or gaps among other parcels and (iii) partial boundary lines which would not create distinct parcels. In mixed parcels where 2D parcels may coexist with 3D parcels (see Fig 7), a specific form of parcel may be needed, described as "liminal"(see section 5.5), ISO19152 (2009), to allow what is in reality a 2D parcel to exactly abut a 3D parcel. Unless there is a complete partitioning of space in 3D, the question of gaps or overlaps is difficult to address. Reasonability checking can assist in locating such errors (e.g. finding very acute angles between adjoining surfaces). See section 5.5 for a discussion on topological encoding.

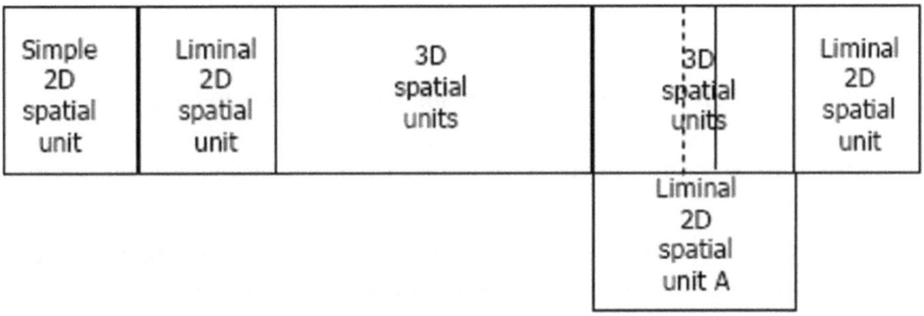

Fig. 7. Top view of mixed 2D and 3D representations ISO19152 (2009)

- *Ensure that the sum of the individual units of 3D floor area per strata was less than or equal to the base 2D parcel.*

This is a trivial and obvious check to verify that the union of all 3D objects per individual strata is totally within the base parcel, where 3D parcel are contained in identifiable strata, such as in a block of units.

- *Ensure that the faces between parcels are complete, so that distinct parcels are formed.*

This situation can arise in 2D when the line dividing two parcels does not connect to the opposite boundary, so that two apparently separate parcels cannot be distinguished (see Fig 7.1). The parcel itself would appear complete, but neighborhood operations within the two intended subdivided parcels would be erroneous. This is a trivial case for a 2D parcel which is treated by most GIS or CAD software, but for a 3D parcel, it becomes a lot more complex to ensure that the faces connect exactly.

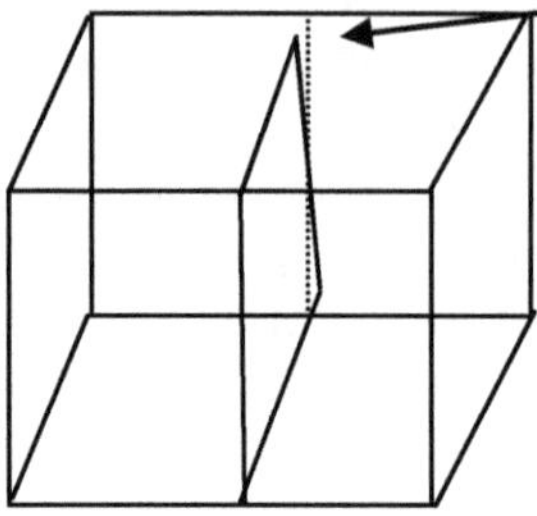

Fig. 7.1. 3D representation of an incomplete subdivision

- *Ensure that self intersections or loops do not create new unintended parcels.*

Again this is not difficult for a 2D parcel and these are already dealt with by most software; however, for a 3D parcel, it is more difficult to ensure that there are no unintended slivers (or shells in 3D), as well as to ensure that once this is resolved, the corrected surfaces remain coplanar (see Fig 7.2).

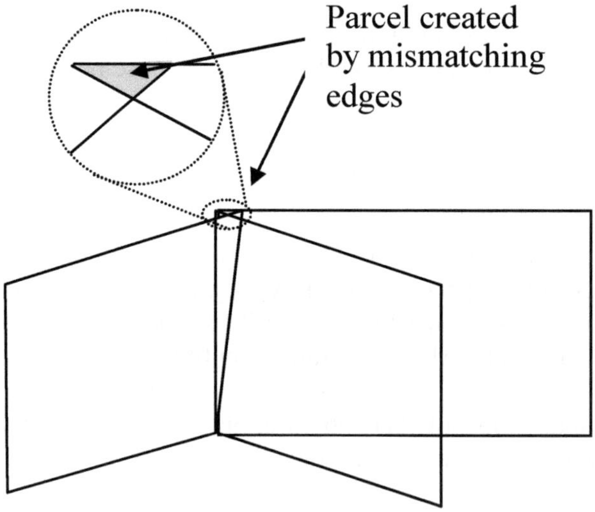

Fig. 7.2. 3D representation of self-intersections creating spurious parcels

4.4. Unique geometrical situations

In this section, unique situations that involve utility networks, or different strata being owned by different parties, existing and officially allowed encroachments or "hanging" parcels creating their own problems have been discussed.

Discussions:

- **Utility Network**. The main complexity of 3D network is its relationship to the base surface parcels (Fig 8). Several issues arise, like:
  - does the network subdivide when surface parcels are subdivided?
  - recording the network when the surface parcel is not mapped in cadastre (e.g. roads, rivers, reserves etc)
  - representing and recording the objects when the networks cross (e.g. tunnels and underground busways),
  - dealing with a situation when the network emerges to the surface,
  - dealing with variable and irregular shapes and sizes etc.

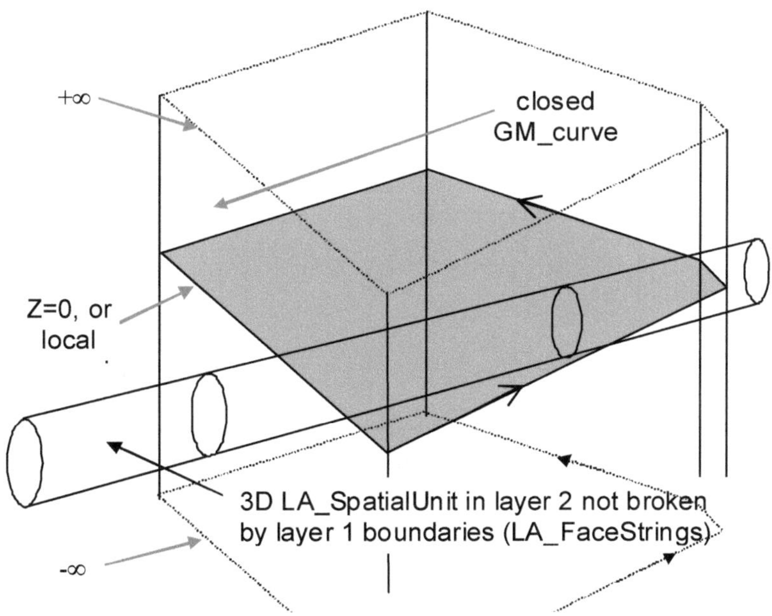

Fig. 8. Utility network object spanning surface parcels (ISO 19152 2009)

- **Multiple strata**. An ideal approach would be to allow multiple owners to rights and restrictions on multiple strata independent of each other, but since most cadastres are based on surface parcels, it becomes complex to maintain a relationship between them. This is highlighted in the case of network objects (see Fig 8) where they will need update operations such as extensions, excisions, splits and merges etc; but as a way around the complexity of maintaining the relationship between the surface and the network object, easements have sometimes been created as subsets of the surface parcels and amalgamated to match the network object. However, if multiple strata titles are allowed (see Figs 8 and 9), constraints must be considered in both the horizontal and vertical direction. In the horizontal direction, the network object would more often than not fall within many surface parcels, which would make it difficult to have a condition where the 3D object lies within the base parcel. In the vertical direction, objects such as crossing networks or tunnels would not encroach in 3D but would in 2D.

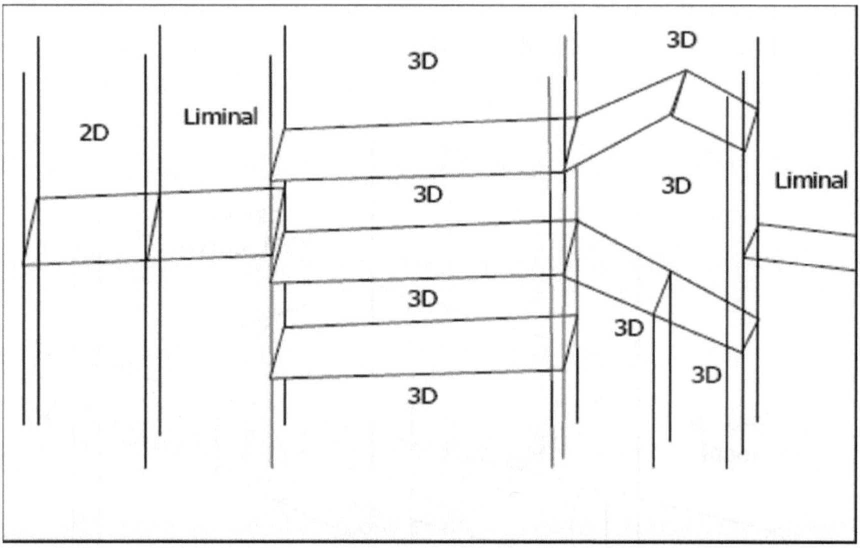

Fig. 9. Mixed use of face string and face to define both bounded and unbounded volumes (ISO 19152 2009). Liminal parcels are 2D parcels which directly adjoin the faces of 3D parcels.

- **Existing encroachment**. If the encroachment is in both the horizontal and vertical directions (2D and 3D), and if 3D volume parcels were created (see Figs 10 and 11), then making sure that the specification allows such parcels to be recorded as legal objects, is quite complex. In such a situation, the validation rules must be able to handle such exceptions as part of the validation process.

Fig. 10. The Gabba Stadium overhanging Stanley Street in Brisbane, Australia

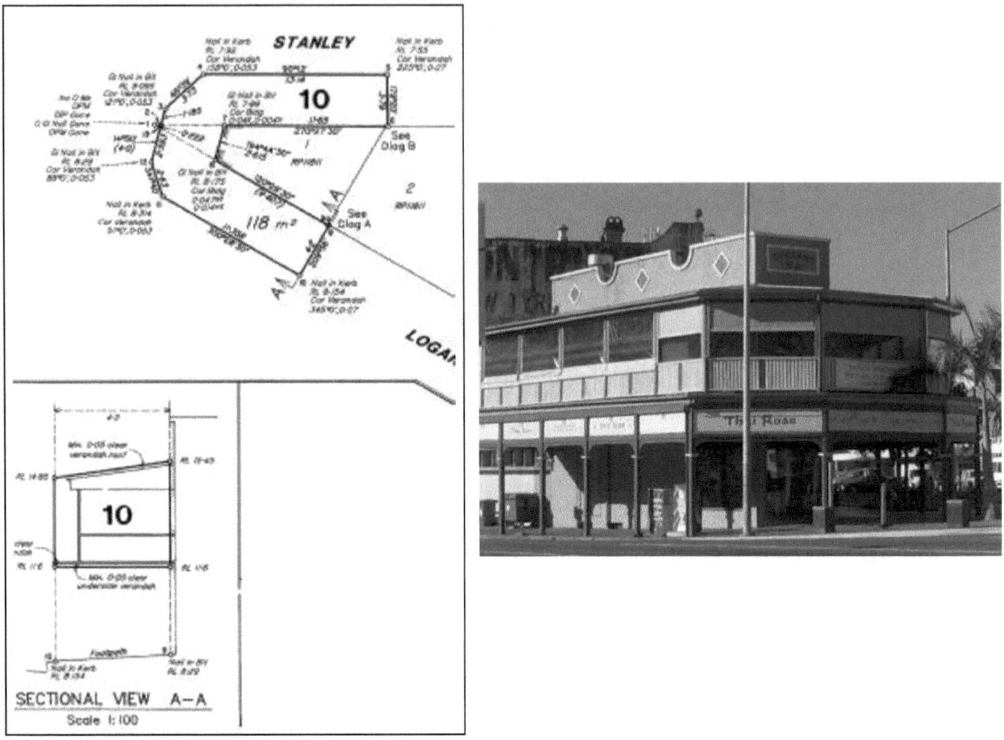

Fig. 11. (Left) 3D parcel number 10 of the Thai Rose, (Right) the restaurant overhanging the footpath on the corner of Stanley Street and Main Street in Brisbane

- **"Hanging" title.** Titles that have no base parcel (see Fig 12), as in the case of the Story Bridge over Brisbane River where one part of the bridge is leased out to permit the lease operators to run a "bridge climb" tour for a fee. This title does not have a base parcel as the bridge is not mapped as a cadastral object, however rights have been created. This creates a complexity in storing and validating such objects as they are not tied to any surface cadastral object.

Fig. 12. The Story Bridge over Brisbane River (Google Earth)

4.5. Further processing of the geometry

***Requirements*:**

1. Effect of subdivision of the 2D base parcel on the 3D parcel and vice versa
2. Consolidation of contiguous or non-contiguous 3D parcels
3. Creating a total/partial reserve/easement on the 3D parcel
4. Limiting of vertical extent of certain 3D parcels by legislation (e.g. in mining areas, owner's rights to the land underground for a group of parcels may be more restricted in depth than other parcels in surrounding areas)

***Discussions*:**

- **Effect of subdivision**

If the 2D parcel is subdivided, it affects the above ground and below ground objects differently. The above ground object such as buildings are also subdivided and it would be just a simple check to see that the 3D object is subdivided proportionately and that the split object is vertical. However, for below ground objects, such as network objects which might extend beyond the parcel, it will be difficult to adapt to circumstances where they might or might not be subdivided. A further complexity is added, if there are two crossing networks below the subdivided parcel or above ground cables with legal space around them for protection, with each network behaving differently as far as subdivision is concerned.

- **Consolidation of 3D parcels**
  Consolidation would be the opposite of subdivision (where the rights, restrictions and responsibilities of two parcels are identical), similar constraints would apply.

- **Creating easements**
  Creating an easement on a 2D base parcel would be similar to subdivision, but if the easement was meant to match a 3D network object and the surface parcel amalgamated, then additional steps would be required to validate the 2D-3D compliance

- **Vertical extents**
  Some areas of the same city, like Ipswich city near Brisbane, can have variable limits to the underground extent based on depth below surface to which owners can exercise their rights, so a reduced level of the ground surface needs to be recorded with the cadastre, so that checks can be made on the extent of the limiting depth.

- **Non Planar Surfaces**
  These greatly increase the complexity of the situation, particularly where curved surfaces intersect, which may result in 3D curved edges with no simple mathematical definition. For this reason, they are usually approximated by a number of planar faces.

4.6. Entry Level Validations

In addition to the above mentioned situations, there may be other circumstances involving geometry which needs to be mentioned for the sake of completeness.

Continuity of parcel.

A group of parcels in a cadastral map may be intersected by, for instance, a road due to development. This will lead to two possible choices: firstly, to maintain each intersected parcels as multi-polygons because they have the same owner and share the same history as the other split part, or, secondly, to create a new record-set for both the divided parcels since the road is an enduring structure, which will permanently divide the parcels and which might even facilitate land transactions for the discrete lots. Both of the cases can be handled by cadastral software, but the choice depends on the predicted practical use of the subsequent data. Validation in such a situation would be limited to a consistent logical approach rather than a rule based method, especially if there is an underground network through the parcel.

Reasonability.

Sometimes when entering data that has been collected in the field or compiled from various sources into the system, due to operator level error, the identification number of two different plans may be incorrectly recorded as the same. This can only be distinguished from the case discussed above if the distance between the parcels exceeds a reasonable value.

Spatio-Temporal.

Although strictly not geometrical, the temporal aspects of cadastre need to be validated as more cadastres round the world are shifting towards a 4D cadastre. Most of the validation in the time dimension will be descriptive and subjective with the spatio-temporal validations being performed at event time itself and thus governed by any of the situations and rules above. In some problem domains it is preferable to treat this as four equivalent as 4D primitives, however, in the cadastral domain the simplifying assumption of changes as atomic events is usually acceptable.

5. Representations of a 3D cadastral parcel as allowed by the ISO 19152 LADM

The ISO19152 LADM, as it is being defined at present, has 5 ways of defining parcels (known in the standard as LA_SpatialUnit). These are:

1. point spatial unit (section 5.1)
2. text spatial unit (section 5.2)
3. line spatial unit (section 5.3)
4. polygon spatial unit (section 5.4)
5. topological spatial unit (section 5.5)

An excerpt of the LADM documentation is included – see Fig 13. Parts of this diagram are used to illustrate the encodings (Figs 14 -19). These encodings are not exclusive and may be mixed within a single database. It is even possible to combine encodings within the definition of the same parcel, e.g. a parcel may be defined by lines on 3 sides but be completed by text.

All of these encodings are applicable in 2D or 3D, and different levels of validation are possible in each case. In general, these are in order of increasing sophistication, but also in order of increasing validation requirements as validation rules will differ for the geometrical/topological primitives and thus the features based on the primitives.

Thus, it can be expected that point based spatial units may be applicable where little detailed knowledge of the cadastre is available because of limited resources. While limited in its analytical functions, this encoding could be used to indicate the location of land parcels against a photographic background.

At the other extreme, the topological encoding allows analytical use of the data, calculation of areas, volumes, determining overlap with other coverages, etc. The cost is that much more stringent validation is needed before data can be included in the database. Correction of these validation errors can be beyond the resources of some authorities. The LADM uses the term "Spatial Unit" for what has been referred to as a parcel. In this section, "Spatial Unit" is used to refer to a parcel.

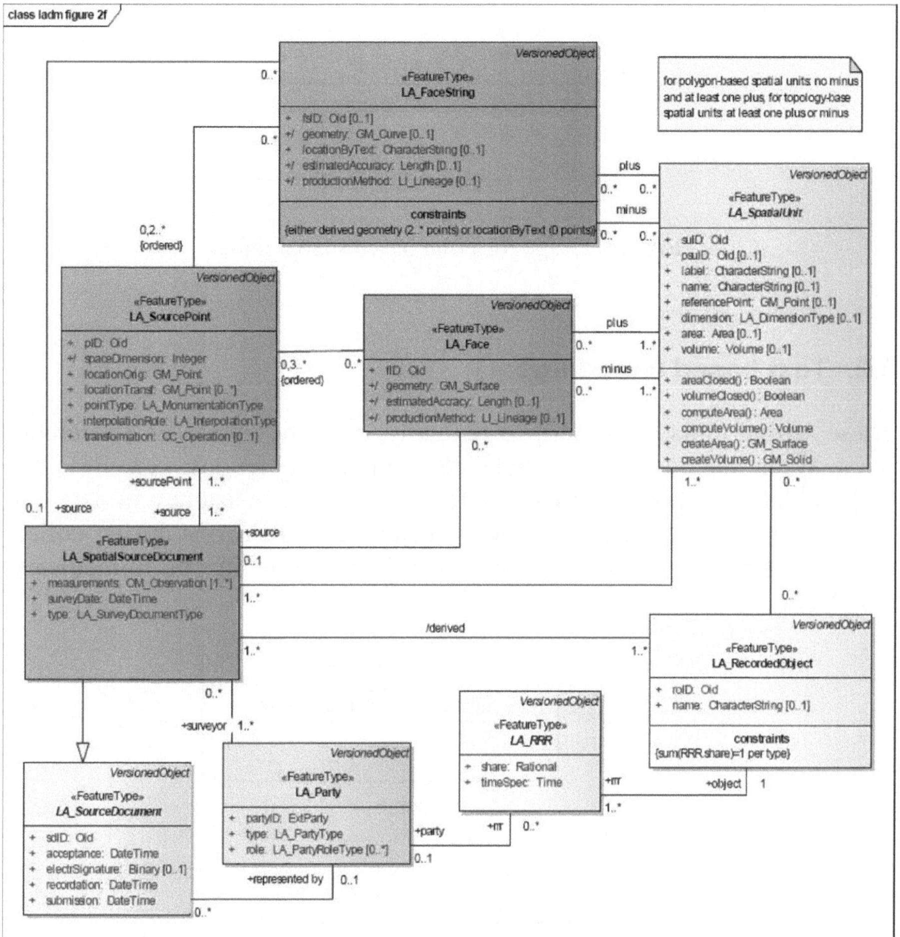

Fig. 13. Classes LA_FaceString, LA_Face, LA_SourcePoint and LA_SpatialSourceDocument from the ISO 19152 LADM being developed

5.1. Point based encodings

Very little validation is possible – except to ensure that the points don't coincide (see Fig 14), and fall within the required jurisdiction

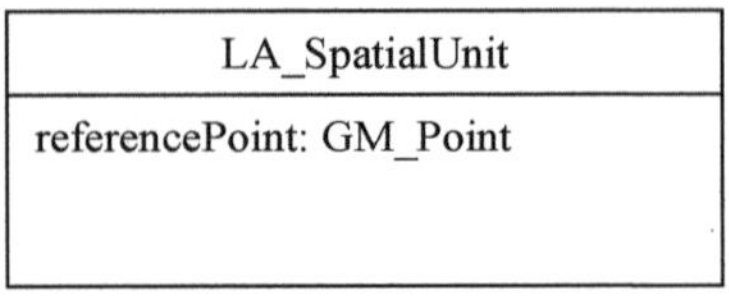

Fig. 14. Usage for Point based spatial units

5.2. Text based encodings

These spatial units are described textually, with the option of a reference point. While the description can be clear and unambiguous, very little formal validation is possible (see Fig 15). However, consistency in data collected and stored can be achieved by using standard forms during collection and storage, such that gross or negligent errors can be immediately visible.

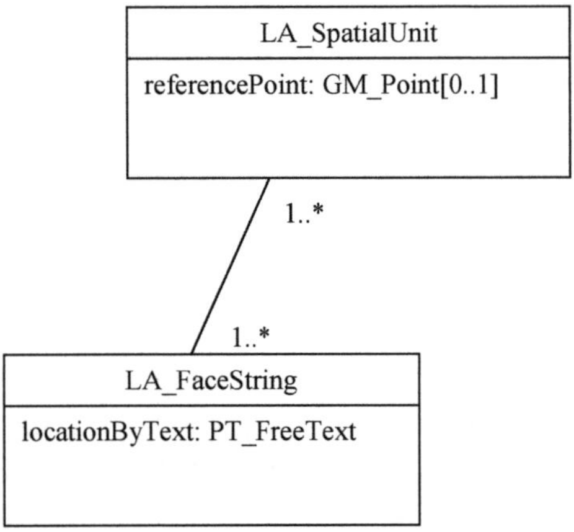

Fig. 15. Usage for Text based spatial units

5.3. Line based encodings

The characteristic of the line-based encoding (otherwise described as 'spaghetti') is that, while validation can be carried out to the same degree as any of the more highly structured approaches; the encoding is capable of interchanging and storing data that contains significant ambiguities, inconsistencies and topological breakdowns. It is suited to the situation where data exists, in a basic and "un-processed" form, but being the best available, must be made available (see Fig 16).

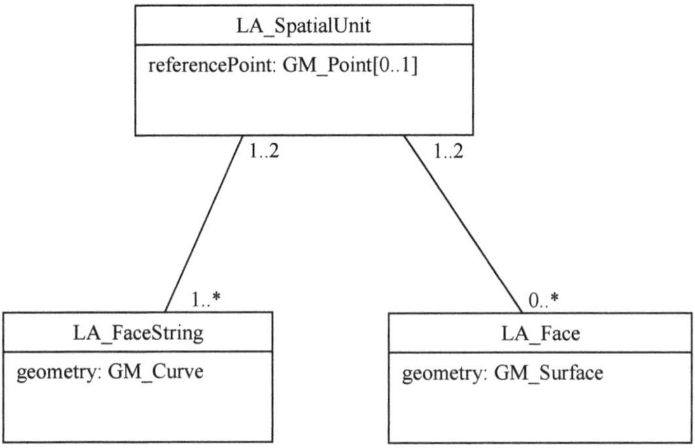

Fig. 16. Usage for Line based spatial units

5.4. Polygon based encodings

The polygon-based encoding stores 2D spatial units as simple polygons, while 3D spatial units are stored as polyhedra. Each FaceString represents one ring, and there should be at least one outer ring and zero or more inner rings. There is no intrinsic constraint that ensures that spatial units do not overlap, or that there are no gaps between them (see Fig 17).

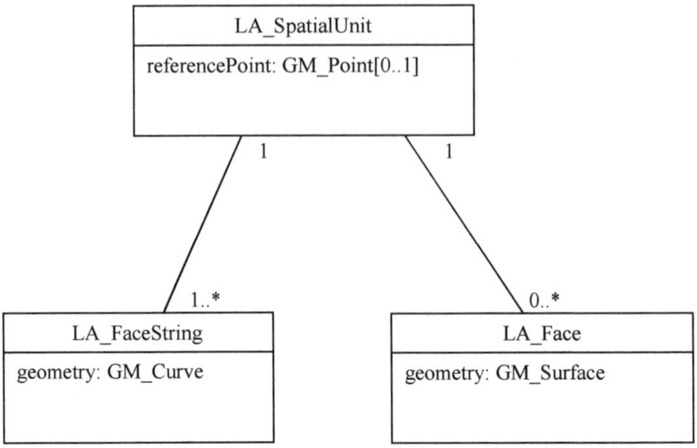

Fig. 17. For polygon-based spatial units

5.5. Topology based encodings

The topologically encoded form of spatial units is similar to the traditional topological encoding, and requires that each faceString, and each face is stored once only, with links to the left and the right spatial units (Figs. 16 and 18). As such, there is an in-built constraint that prevents certain kinds of invalid data from being represented, and is particularly suited to the situation where a complete non-overlapping partition of space is to be defined.

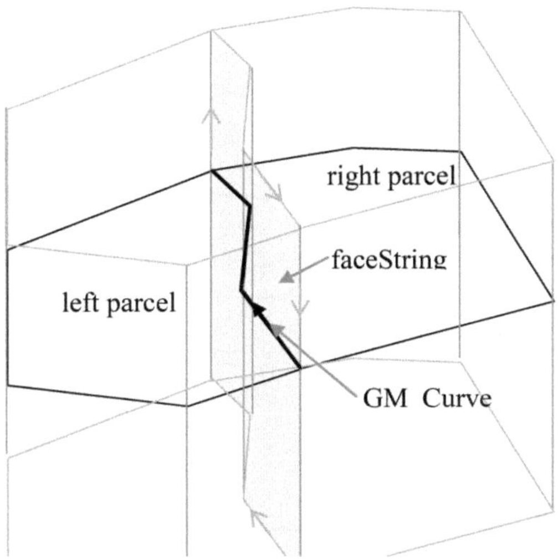

Fig. 18. Topologically encoded 2D spatial units (with common faceString shaded) (ISO19152 2009)

The combinations of 2D and 3D spatial units that adjoin, create an extra complexity for the topological encoding form (see Fig 19). Because a 2D spatial unit may directly adjoin a 2D spatial unit, an additional type of spatial unit (described as "liminal") is needed. This parcel is intrinsically a 2D parcel, but has to be represented with faces in its boundary. (ISO19152 2009).

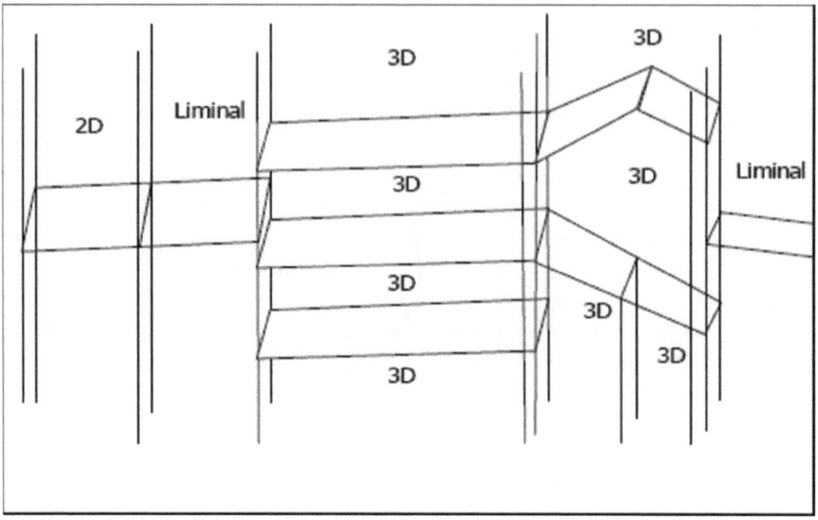

Fig. 19. Reprint of Fig 9, showing topological encoding of the junction of 2D and 3D spatial units, side view of Fig 7 (ISO19152 2009)

6. Discussion and future work

The amount and completeness of the validation that needs to be applied to cadastral data varies according to the level of maturity and sophistication of the databases and applications involved. For that reason, ISO19152 defines alternative encoding strategies to ensure that for whatever level of validation is achievable, the data can still be represented.

The decision as to what validation rules are to be applied by a particular registering authority will be made by that authority, and should be enabled by technology. This philosophy will ensure that authorities are not inhibited from developing a cadastral database by the high cost of ensuring pristine data.

As an example, a cadastre may begin life with a mixture of point and text based spatial units, gradually being upgraded to a line-based encoding. This form can be shown to be useful for certain applications, in spite of topological failings. Further effort could then result in a fully topological cadastral database.

It is significant that all five levels of maturity that are accommodated by ISO19152 can be applied in 2D and 3D, and even more importantly, to a combination of 2D and 3D parcels.

6.1. Further Research

This paper provides a preliminary investigation of the issues concerned with validation. It raises questions and identifies problems that need to be resolved to ensure a functional 3D cadastral system. As a generic study, in-depth analysis and investigations can be carried out on each aspect of the validation requirements during the stages of data collection, storage and processing. As specific requirements will differ, and not all of the issues raised in this paper are relevant to all 3D cadastres, user specific validation requirements need to be studied to suit individual circumstances.

7. Bibliography

ESRI, (2004) Parcels and the Cadastre Data Model Poster, Land Parcels Data Model, Accessed May 11, 2009 from http://support.esri.com/index.cfm?fa=downloads.dataModels.gateway.

Friedrich J, (2004) Spatial Modeling in Natural Sciences and Engineering, Springer, Berlin, ISBN 3-540-20877-1.

ISO19107: ISO-TC211 (2001). Geographic Information - Spatial Schema. Geneva, International Organization for Standards.

ISO 19152: ISO-TC211 (2009), Geographic Information, Land Administration Domain Model (LADM) Committee Draft Version 1, International Organization for Standardization Technical Committee 211 distributed within ISO and CEN for voting 10 July 2009.

Kolbe TH (2009) Representing and Exchanging 3D City Models with CityGML. In: 3D Geo-Information Sciences. Lee J, Zlatanova S (Eds). Springer, Berlin.

Ledoux H, Meijers M (2009) Extruding building footprints to create topologically consistent 3D city models, In: Urban and Regional Data Management, UDMS Annual 2009, Krek A, Rumor M, Zlatanova S, Fendel E (Eds). CRC Press.

Lemmen C, Oosterom P, Uitermark H, Thompson R, Hespanha J (2009) Transforming the Land Administration Domain Model into an ISO standard (LADM as ISO 19152), FIG Working Week, Eilat, Israel.

Kazar BM, Kothuri R, van Oosterom P, Ravada S (2008) On Valid and Invalid Three-Dimensional Geometries. In: Advances in 3D Geoinformation Systems. Van Oosterom P, Penninga F, Zlatanova S, Fendel E (Eds). Springer, Berlin.

OGC (1999) Open GIS Simple features Specification for SQL, Revision 1.1. Retrieved Aug 10, 2009, from http://www.opengeospatial.org/standards/sfs.

OGC. (2004). "Geography Markup Language (GML)." GML-3.1.0.doc Retrieved Mar 2007, from http://www.opengeospatial.org/standards/gml.

Van Oosterom P, Zlatanova S (2008) Creating Spatial Information Infrastructures, CRC Press, Taylor and Francis Group, Boca Raton, USA, ISBN 978-1-4200-7068-2.

Penninga F, Van Oosterom P, Kazar BM (2006) A TEN-based DBMS Approach for 3D Topographic Data Modeling. Spatial Data Handling 06. 12th International Symposium on Spatial Data Handling, Springer, http://www.rgi-otb.nl/3dtopo/documents/sdh06_TEN_dbms_v7_final3.pdf

Rahman AA, Pilouk M (2007) Spatial Data Modelling for 3D GIS, Springer, New York, ISBN 978-3-540-74166-4.

Stoter J (2004) 3D Cadastre. PhD thesis, Delft University of Technology, The Netherlands. http://www.ncg.knaw.nl/Publicaties/Geodesy/pdf/57Stoter.pdf

Stoter J, Oosterom P (2006) 3D Cadastre in an International Context, CRC Press, Taylor and Francis Group, Boca Raton, USA, ISBN 0-8493-3932-4.

Thompson RJ (2007) Towards a rigorous logic for spatial data representation, PhD thesis, Delft University of Technology, The Netherlands, http://www.rgi-otb.nl/3dtopo/documents/RGI-011-52.pdf

Thompson RJ, Van Oosterom P (2007) Mathematically provable correct implementation of integrated 2D and 3D representations. In: Advances in 3D Geoinformation Systems. Van Oosterom P, Penninga F, Zlatanova S, Fendel E (eds). Springer, Berlin.

From Three-Dimensional Topological Relations to Contact Relations

Yohei Kurata

SFB/TR8 Spatial Cognition, Universität Bremen
Postfach 330 440, 28334 Bremen, Germany

Abstract. Topological relations, which concern how two objects intersect, are one of the most fundamental and well-studied spatial relations. However, in reality, two physical objects can take only *disjoint* relation if they are solid. Thus, we propose an alternative of topological relations, called *contact relations*, which capture how two objects contact each other. Following the framework of the 9-intersection, this model distinguishes contact relations based on the presence or absence of contacts between several surface elements of two objects. Consequently, the contact relations have a strong correspondence to the 9-intersection-based topological relations. Making use of this correspondence, we derive the contact relations between various combinations of objects in a 3D Euclidean space ($\mathbb{R}^3$). For this purpose, we first review and analyze the topological relations in $\mathbb{R}^3$. Then, these topological relations are mapped to contact relations.

1. Introduction

Qualitative spatial relations are fundamentals for characterizing spatial arrangements of objects in a space, usually following how people conceptualize them. Spatial database communities have studied several types of qualitative spatial relations, among which topological relations have attracted much attention due to its importance in human spatial cognition (Shariff, et al. 1998). Egenhofer (1989), for instance, distinguished eight types of topological relations between two regions in a 2D Euclidean space

T. Neutens, P. De Maeyer (eds.), *Developments in 3D Geo-Information Sciences*,
Lecture Notes in Geoinformation and Cartography, DOI 10.1007/978-3-642-04791-6_7,

$\mathbb{R}^2$, called *disjoint*, *meet*, *overlap*, *covers*, *coveredBy*, *contains*, *inside*, and *equal* relations. The same set of eight relations holds between two bodies in $\mathbb{R}^3$ (Zlatanova 2000). Similarly, Randell *et al.* (1992) distinguished eight types of topological relations without regard to the dimension of the underlying space. Such topological relations, however, fail when they are applied to physical objects—simply because they are usually solid and, accordingly, they do not intersect/connect with each other (note that the connection of two objects in Randell *et al.*'s sense is established by the presence of a common point, thereby different from the notion of contact).

Although the solid objects are always topologically disjoint (disconnected) in theory, people often care about whether they are totally separated or having a contact and, if they have a contact, in what way. For instance, we use the expressions like *the bottom of A meets the top of B* (or simply *A is on B*, although this expression has more broad meanings (Feist 2000)) or *A stands on the edge/corner of B* (Billen and Kurata 2008). These expressions concern which part of one object contacts which part of another object. Such information is critical when we characterize the spatial arrangement of two contacting objects. We, therefore, propose a new model of qualitative spatial relations that captures how two objects contact each other, called *contact relations*.

Contact may occur only between the surfaces of two objects. In a 3D space, for each region, we can distinguish its front and back sides. Similarly, we can distinguish several sides of bodies considering their shapes (cubes, pyramids, etc.). Thus, the model of contact relations should be able to record the presence or absence of contacts between several surface elements of two objects in different granularities. Our model, therefore, follows the frameworks of the *9-intersection* (Egenhofer and Herring 1991) and the *9⁺-intersection* (Kurata and Egenhofer 2007), which distinguish topological relations based on the presence or absence of intersections between the topological parts of two objects. Naturally, contact relations in our model have a strong correspondence with 9-intersection-based topological relations. Indeed, making use of this correspondence, we derive contact relations between various combinations of objects (Section 5).

The remainder of this paper is organized as follows: Section 2 summarizes the concepts of the 9- and 9⁺-intersection. Based on these models, Section 3 develops a model of contact relations. Section 4 reviews topological relations in $\mathbb{R}^3$ and discusses the property of these relations for later discussions. Section 5 derives six sets of contact relations in $\mathbb{R}^3$ from the corresponding sets of topological relations discussed in Section 4. Finally, Section 6 concludes with a discussion of future problems.

In this paper, *solid objects* refer to the objects that do not intersect. We can consider that any sort of objects (points, lines, regions, and bodies) may be solid or non-solid objects. Our target is limited to solid objects. In addition, to simplify, we consider only simple objects (Schneider and Behr 2006). Simple objects consist of a single connected component. Simple lines are non-branching lines without loops, and simple regions/bodies are regions/bodies without holes, spikes, cuts, and disconnected interior.

2. The 9-Intrersection and The 9⁺-Intersection

Our model of contact relations follows the 9-intersection (Egenhofer and Herring 1991) and its refinement called the 9^{+}-intersection (Kurata and Egenhofer 2007). Based on point-set topology (Alexandroff 1961), these two models distinguish the *interior*, *boundary*, and *exterior* each object. The interior of an object X, denoted X°, is the union of all open sets contained in X, X's boundary ∂X is the difference between X's closure (the intersection of all closed point sets that contain X) and X°, and X's exterior X^{-} is the complement of X's closure. By definition, the boundary of a line refers to the set of its two endpoints, and a point does not have an interior. In the 9-intersection, topological relations are characterized by the properties of the 3×3 intersections between the topological parts (interior, boundary, and exterior) of two objects. These 3×3 intersections are concisely represented in the *9-intersection matrix* in Eq. 1. In the most basic approach, topological relations are distinguished by the presence or absence of these 3×3 intersections. Accordingly, topological relations are characterized by the pattern of bitmap-like icons (Mark and Egenhofer 1994), whose 3×3 cells indicate the emptiness/non-emptiness of the 3×3 elements in the 9-intersection matrix (Fig. 1).

$$\mathrm{M}(A,B) = \begin{pmatrix} A^{\circ} \cap B^{\circ} & A^{\circ} \cap \partial B & A^{\circ} \cap B^{-} \\ \partial A \cap B^{\circ} & \partial A \cap \partial B & \partial A \cap B^{-} \\ A^{-} \cap B^{\circ} & A^{-} \cap \partial B & A^{-} \cap B^{-} \end{pmatrix} \quad (1)$$

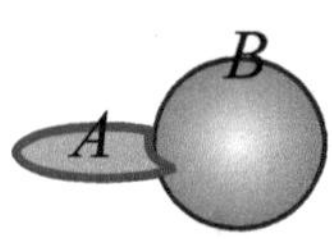

$$\mathrm{M}(A,B) = \begin{pmatrix} \emptyset & \emptyset & \neg\emptyset \\ \emptyset & \neg\emptyset & \neg\emptyset \\ \neg\emptyset & \neg\emptyset & \neg\emptyset \end{pmatrix} \rightarrow$$

Fig. 1. Representations of a topological relation by the 9-intersection matrix/icon

The 9^{+}-intersection extends the 9-intersection by considering the *subdivision* of topological parts. For instance, the boundary of a line can be subdivided into two endpoints. To support such subdivision, the nested version of the 9-intersection matrix shown in Eq. 2, called the *9⁺-intersection matrix*, is used, where $A^{\circ i}$, $\partial_i A$, and A^{-i} are the ith subpart of *A*'s interior, boundary, and exterior, while $B^{\circ j}$, $\partial_j B$, and B^{-j} are the jth subpart of *B*'s interior, boundary, and exterior, respectively. Just like the 9-intersection, topological relations are distinguished by the presence or absence of all intersections listed in the 9^+-intersection matrix. For instance, Fig. 2 shows the 9^+-intersection matrix and its iconic representation of a topological relation between a (directed) line and a region. To simplify, the bracket of each matrix element is omitted if it consists of a single sub-element.

$$M^+(A,B) = \begin{pmatrix} [A^{\circ i} \cap B^{\circ j}] & [A^{\circ i} \cap \partial_j B] & [A^{\circ i} \cap B^{-j}] \\ [\partial_i A \cap B^{\circ j}] & [\partial_i A \cap \partial_j B] & [\partial_i A \cap B^{-j}] \\ [A^{-i} \cap B^{\circ j}] & [A^{-i} \cap \partial_j B] & [A^{-i} \cap B^{-j}] \end{pmatrix} \quad (2)$$

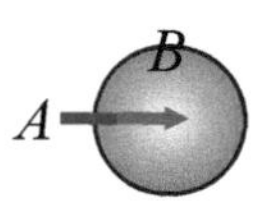

$$M^+(A,B) = \begin{pmatrix} A^\circ \cap B^\circ & A^\circ \cap \partial B & A^\circ \cap B^- \\ \begin{bmatrix} \partial_1 A \cap B^\circ \\ \partial_2 A \cap B^\circ \end{bmatrix} & \begin{bmatrix} \partial_1 A \cap \partial B \\ \partial_2 A \cap \partial B \end{bmatrix} & \begin{bmatrix} \partial_1 A \cap B^- \\ \partial_2 A \cap B^- \end{bmatrix} \\ A^- \cap B^\circ & A^- \cap \partial B & A^- \cap B^- \end{pmatrix} = \begin{pmatrix} \neg\emptyset & \neg\emptyset & \neg\emptyset \\ \begin{bmatrix} \emptyset \\ \neg\emptyset \end{bmatrix} & \begin{bmatrix} \emptyset \\ \emptyset \end{bmatrix} & \begin{bmatrix} \neg\emptyset \\ \emptyset \end{bmatrix} \\ \neg\emptyset & \neg\emptyset & \neg\emptyset \end{pmatrix} \rightarrow$$

Fig. 2. Representations of a topological relation by the 9^+-intersection matrix/icon

3. Modeling Contact Relations

Contact relations between two objects are distinguished basically based on the presence or absence of contacts between their surface elements. Surface elements of an object are the parts of the object that is connected directly to the object's exterior. For regions in $\mathbb{R}^3$, we can naturally distinguish three surface elements; the front side, the back side, and the edge. The front and back sides correspond to the region's topological interior, while the edge corresponds to the region's topological boundary. For bodies, we may distinguish different numbers of surface elements; for instance, we can distinguish six sides of a cubic body. To simplify, however, this paper does not distinguish the sides of bodies and considers that the body has only one surface element, which corresponds to the body's topological boundary. For lines, we consider that they have two surface elements, a set of two endpoints and an intermediate segment, which correspond to the line's topological boundary and interior, respectively.

In this way, surface elements of objects are distinguished in association with the topological part of the objects. Making use of this association, we use a 9^+-intersection-like matrix, called the *9^+-contact matrix* (Eq. 3), for representing contact relations. In this matrix, each Boolean function $\mathrm{c}(X, Y)$ indicates whether X contacts Y or not. X and Y may refer to a surface element of an object, which we define as a subpart of the object's interior or boundary. X and Y may also refer to the exterior of an object. By recording the presence or absence of contact between A's surface elements (say X) and B's exterior B^-, we can tell from the matrix whether X contacts the surface of another object entirely, partly, or not at all—for instance, if X contacts only one of B's surface element Y and not B^-, X entirely contacts Y. The most right-bottom element in the 9^+-contact matrix is fixed, as it does not contribute to the distinction of contact relations. This does not necessarily mean that we should interpret that the exteriors of two objects contact each other.

$$\mathrm{C}^+(A,B) = \begin{pmatrix} [\mathrm{c}(A^{\circ_i}, B^{\circ_j})] & [\mathrm{c}(A^{\circ_i}, \partial_j \mathrm{B})] & [\mathrm{c}(A^{\circ_i}, B^-)] \\ [\mathrm{c}(\partial_i A, B^{\circ_j})] & [\mathrm{c}(\partial_i A, \partial_j \mathrm{B})] & [\mathrm{c}(\partial_i A, B^-)] \\ [\mathrm{c}(A^-, B^{\circ_j})] & [\mathrm{c}(A^-, \partial_j \mathrm{B})] & true \end{pmatrix} \quad (3)$$

Since all elements in the 9^+-contact matrix are two-valued (*true* or *false*), the patterns of the 9^+-contact matrix are represented by bitmap-like icons (Fig. 3) in which each black/white cell indicates the true/false value of the corresponding element in the 9^+-contact matrix. In the example of Fig. 3 the icon's first column and first row are partitioned, since they correspond to the front and back sides of the regions.

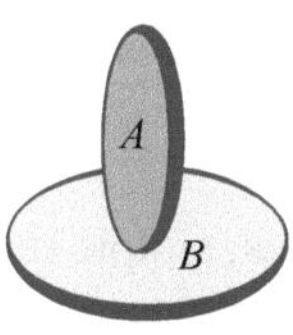

$$\mathrm{C}^+(A,B) = \begin{pmatrix} \begin{bmatrix} false & false \\ false & false \end{bmatrix} & \begin{bmatrix} false \\ false \end{bmatrix} & \begin{bmatrix} true \\ true \end{bmatrix} \\ [true \;\; false] & false & true \\ [true \;\; true] & true & true \end{pmatrix} \rightarrow$$

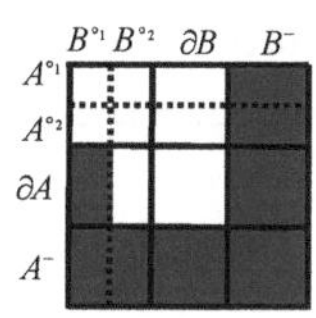

Fig. 3. Representations of a contact relation by the 9^+-contact matrix/icon

An attention should be paid to the definition of the contact between the boundary (edge) of a region and the exterior of another object. We say that the boundary of a region A contacts the exterior of an object B if and only if A's boundary has an interval that does not contact B's surface. Accordingly, in Fig. 4, we consider that A's edge does not contact B's exterior.

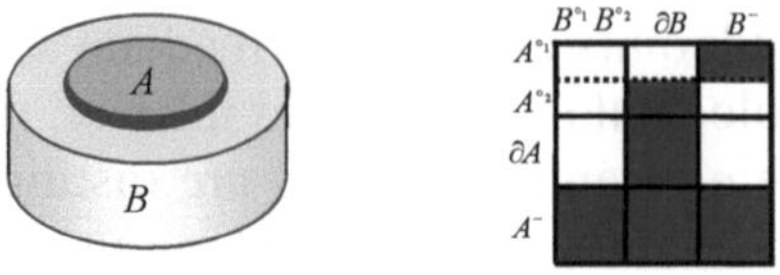

Fig. 4. An example case where we consider that a region's boundary does not contact a body's exterior

Dimensional model by Billen *et al.* (2002) also considers the distinction of surface elements (corner points, edges, and faces) for distinguishing *dimensional relations*, which are essentially a refinement of topological relations. Later Billen and Kurata (2008) reformulated the dimensional model using the framework of the 9^+-intersection, giving a new name *projective 9^+-intersection*. Our model is structurally similar to this projective 9^+-intersection, although our model targets the contacts while the projective 9^+-intersection deals with the intersections. In addition, the front and back sides of regions are not distinguished in the projective 9^+-intersection.

4. Three-Dimensional Topological Relations

This section reviews and analyzes topological relations in $\mathbb{R}^3$, which are later used for deriving contact relations. The topological relations in $\mathbb{R}^3$ were studied by Zlatanova (2000). Later, Kurata (2008) systematically re-identified the sets of topological relations between various combinations of objects in $\mathbb{R}^1$, $\mathbb{R}^2$, and $\mathbb{R}^3$, making use of the universal constraints on the 9^+-intersection. Table 1 shows the numbers of the identified relations. Among these relation sets, this paper uses point-body relations (Fig. 5), line-region relations (Fig. 6), line-body relations (Fig. 7), region-region relations (Fig. 8), region-body relations (Fig. 7), and body-body relations (Fig. 9), each of which contains the relations that hold only in $\mathbb{R}^3$.

Table 1. Numbers of topological relations between points, simple lines, simple regions, and simple bodies in $\mathbb{R}^1$, $\mathbb{R}^2$, and $\mathbb{R}^3$, distinguished by the emptiness/non-emptiness patterns of the 9-intersection (Kurata 2008)

	$\mathbb{R}^1$	$\mathbb{R}^2$	$\mathbb{R}^3$		$\mathbb{R}^1$	$\mathbb{R}^2$	$\mathbb{R}^3$
Point-Point	2	2	2	Line-Region	–	19	31
Point-Line	3	3	3	Line-Body	–	–	19
Point-Region	–	3	3	Region-Region	–	8	44
Point-Body	–	–	3	Region-Body	–	–	19
Line-Line	8	33	33	Body-Body	–	–	8

4.1. Correspondence between Topological Relation Sets

Comparison of the relation sets in Figs. 5-9 reveals several correspondences between the relation sets. The most remarkable correspondence is that between the 19 line-body relations and the 19 region-body relations, which are represented by almost the same sets of icons (Fig. 7). The 19 line-body relations also correspond to the 19 of the 31 line-region relations in Figs. 6a-s. Note that these 19 line-region relations are the relations that may hold also in $\mathbb{R}^2$. Considering that both the boundary of a body and that of a region forms a Jordan curve in $\mathbb{R}^3$ and $\mathbb{R}^2$, respectively, the correspondence between the line-body relations in $\mathbb{R}^3$ and the line-region relations in $\mathbb{R}^2$ is reasonable. For the same reason, the 8 body-body relations (Fig. 9) correspond to the 8 of the 43 region-region relations in Figs. 8a-$f_{1/2}$, which may hold also in $\mathbb{R}^2$.

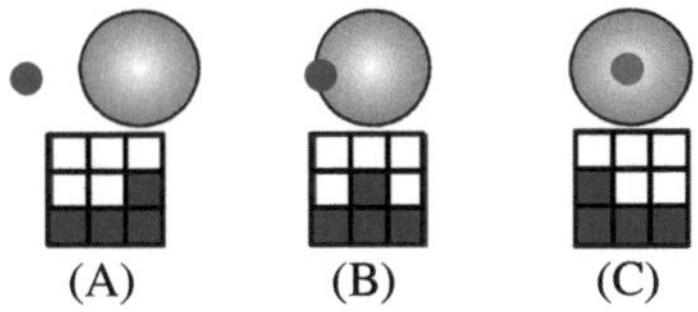

Fig. 5. 3 topological point-body relations in $\mathbb{R}^3$

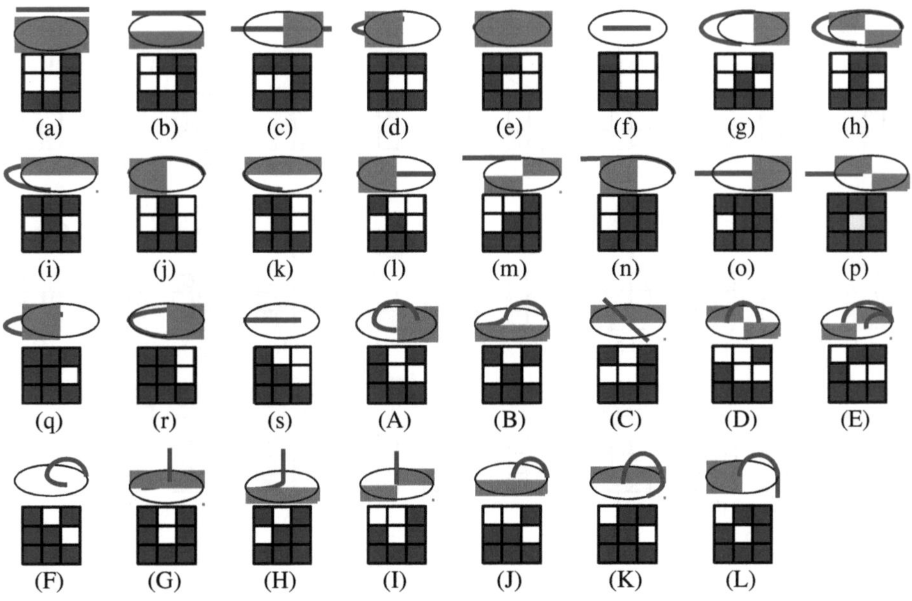

Fig. 6. 31 topological line-region relations in $\mathbb{R}^3$

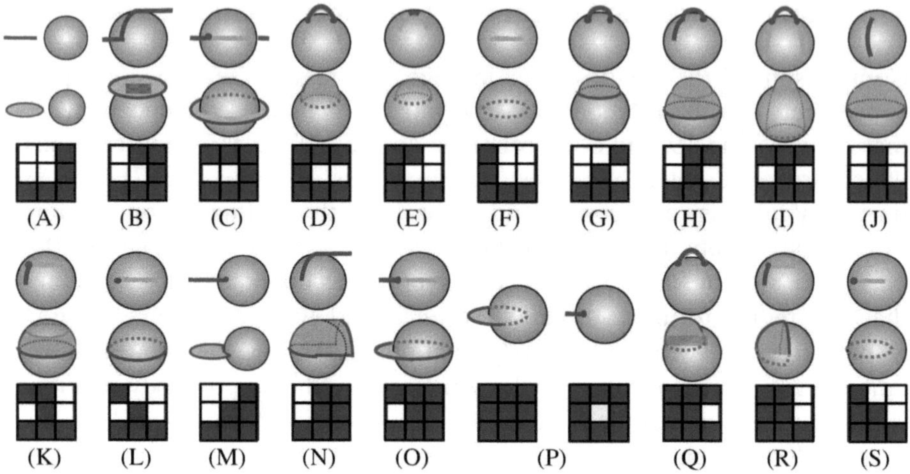

Fig. 7. 19 topological line-body and region-body relations in $\mathbb{R}^3$

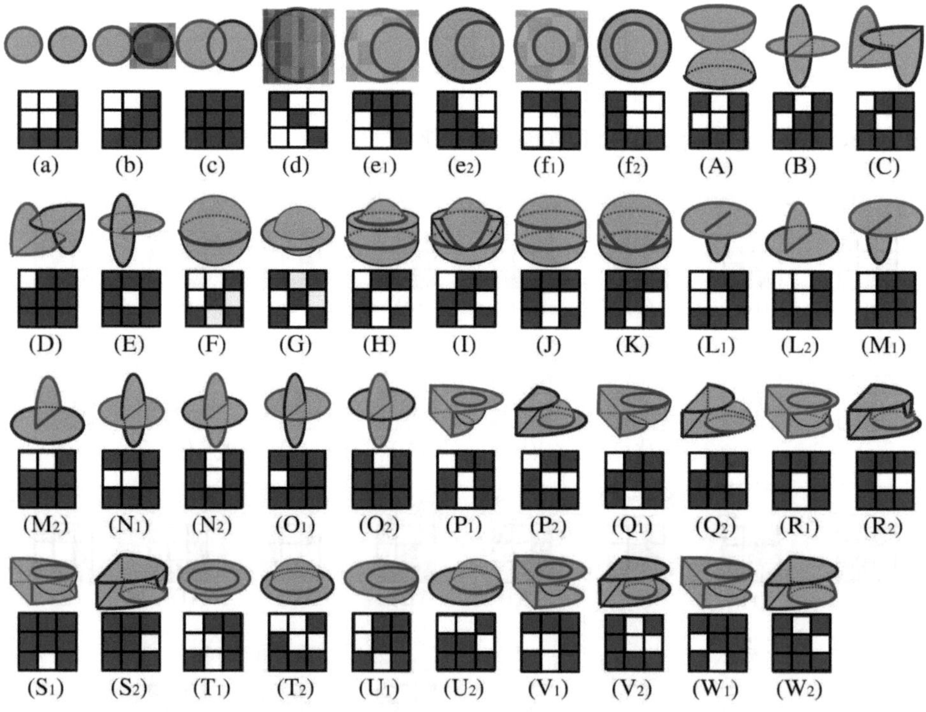

Fig. 8. 43 topological region-region relations in $\mathbb{R}^3$

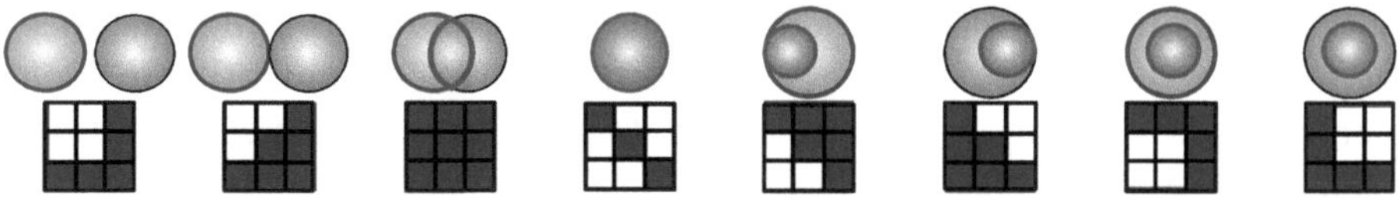

(A) (B) (C) (D) (E_1) (E_2) (F_1) (F_2)

Fig. 9. 8 topological body-body relations in $\mathbb{R}^3$

4.2. Topological Relations That Hold between Convex Objects

For each topological relation set, we identify the relations that hold even when the objects are limited to convex ones. For instance, the line-region relations in Figs. 8a-$f_{1/2}$ hold between two convex regions, while the relation in Fig. 8A does not. The reason why we pay attention to such relations between convex objects, since convex objects are *prototypical*—convex lines are straight and convex regions are flat—and also many physical objects in the real world are convex.

The topological relations that cannot hold between two convex objects (say, *A* and *B*) are filtered out by the following constraints:

- if *A*'s boundary intersects with *B*'s interior only or with *B*'s interior and boundary, but not *B*'s exterior, then *A*'s interior intersects with *B*'s interior only (and vice versa);
- if *A*'s boundary intersects with *B*'s boundary only, then *A*'s interior intersects with *B*'s interior only or *B*'s boundary only (and vice versa);
- *A*'s interior and *B*'s interior intersect if and only if *A*'s boundary intersects with *B*'s closure or *B*'s boundary intersects with *A*'s closure.

By sketching an instance manually it is confirmed that the remaining topological relations hold between two convex objects. In this way, we identified the following topological relations that hold even when the objects are limited to convex ones:

- all point-body relation (Figs. 5A-C);
- 13 line-region relations in Figs. 6a-c,f,j,l-p,s,C,I;
- 11 line-body relations in Figs. 7A-C,F,J,L-P,S;
- 11 region-body relations in Figs. 7A-C,F,J,L-P,S;
- 18 region-region relations in Figs. 8a-$f_{1/2}$,,B,$L_{1/2}$-$O_{1/2}$; and
- all body-body relations (Figs. 9A-$F_{1/2}$).

4.3. Topological Relations with Piercing

We say that an object *A* pierces an object *B* when *A*'s interior intersects continuously with *B*'s interior and its neighboring exterior on two different sides. In our framework, the pierced object *B* is limited to a region, since the sides are defined only for regions (Section 3). We pay attention to such piercing because topological relations that require piercing cannot be mapped to contact relations (Sections 5.3-5.4). Piercing are found in the sample configurations of the relations in Fig. 6C, Figs. 7C-F,I,K,L,O-S, and Figs. 8B,E,$N_{1/2}$-$O_{1/2}$. This does not mean that these relations require piecing. In fact, Fig. 10 shows that the line-region relation in Fig. 6C and the region-region relations in Figs. 8B,E,$N_{1/2}$-$O_{1/2}$ do not require piercing (when two objects are limited to convex ones, however, these topological relations require piercing). On the other hand, the region-body relations in Figs. 7C-F,I,K,L,O-S always require piercing, because whenever the body's interior intersects with the region's interior, the body's interior also intersects continuously with the region's neighboring exterior on both sides.

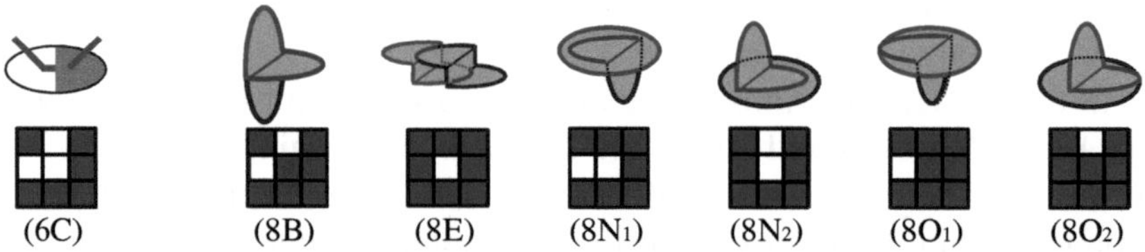

Fig. 10. Alternative configurations of the line-region relation in Fig. 6C and the region-region relations in Figs. 8B,E,$O_{1/2}$-$M_{1/2}$, which do not have piercing.

4.4. Topological Relations with Enclosed Spaces

In Figs. 8F-K,$P_{1/2}$-$W_{1/2}$, we can find a space (or spaces) enclosed by two regions. Actually, any configuration of the region-region relations in Figs. 8F-K,$P_{1/2}$-$W_{1/2}$ have enclosed space(s), because the closure of one region contains the boundary of another region, but not its interior. Similarly, any configuration of the region-body relations in Figs. 7D,G,H,Q have enclosed space(s). The presence of such enclosed space imposes some restrictions on the corresponding contact relations (Section 5.4).

5. Deriving Contact Relations

Contact relations have a strong correspondence with topological relations, as we can imagine from the structural similarity between the 9^+-contact matrix and the 9-/9^+-intersection matrix. Normally, contact relations can be derived from topological relations simply by replacing intersections with contacts (Figs. 11a-b). As for topological relations between a region and an object *X*, each topological relation is potentially mapped to multiple contact relations, since the intersection between the region's interior and *X*'s interior/boundary is mapped to contacts between either or both faces of the region and *X*'s interior/boundary (Fig. 11b). On the other hand, there are some topological relations that cannot be mapped to contact relations , such as those with piercing (Fig. 11c). In the following subsections, we derive the sets of contact relations between various combinations of objects based on the mapping from the corresponding sets of topological relations.

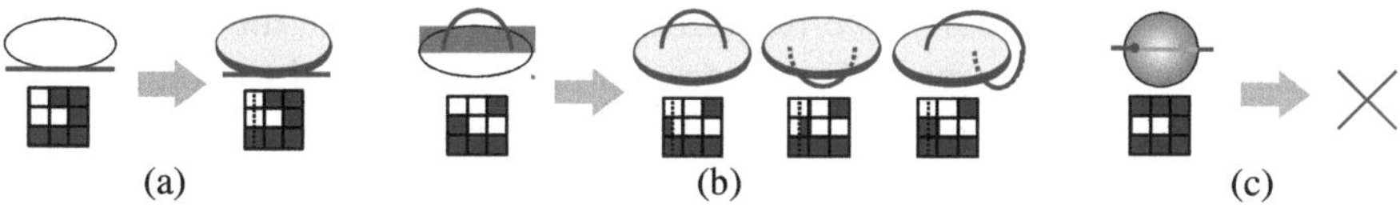

Fig. 11. Sample mappings from topological relations to contact relations

5.1. Point-Body, Line-Body, and Body-Body Relations

The topological point-body relations in Fig. 5A,B are mapped to one contact relation, respectively, while the relation in Fig. 5C cannot be mapped to contact relations because the point is inside the body. Consequently, two contact relations between a point and a body in $\mathbb{R}^3$, represented by the 9^+-contact icons  , are derived. In a similar way, seven contact relations between a line and body in $\mathbb{R}^3$,       , are derived from the topological line-body relations in Figs. 7A,B,G,H,J,M,N, and two contact relations between two bodies in $\mathbb{R}^3$,  , are derived from the topological body-body relations in Figs. 9A,B.

Among these contact relations, which hold when the objects are limited to convex ones? Section 4.2 found that the topological point-body relations in Fig. A,B, the topological line-body relations in Figs. 7A,B,J,M,N, and the topological body-body relations in Figs. 9A,B hold in convex cases. Naturally, the contact relations derived from these topological relations are the relations that hold in convex cases; i.e.,

- two contact relations,  , hold between a point and a convex body;
- five contact relations,     , hold between a convex line and convex body; and
- two contact relations,  , hold between two convex bodies.

5.2. Region-Body Relations

Contact relations between a region and a body in $\mathbb{R}^3$ are derived from the 19 topological region-body relations in Fig. 7. The relations in Figs. 7A,G,M are mapped to one contact relation, respectively. Accordingly, 3×1 contact relations, 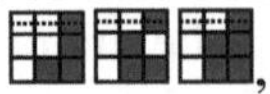, are derived from them. The relations in Figs. 7B,H,J,N are mapped to three contact relations, respectively, because the body's boundary contacts the region's either or both sides. Thus, 4×3 contact relations, 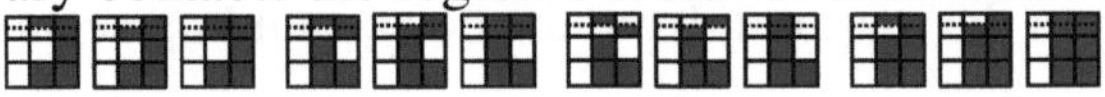, are derived from them. Finally, the remaining relations, where the body's interior intersects with the region's closure, cannot be mapped to contact relations. In this way, in total 3×1+4×3 = 15 contact relations between a region and a body in $\mathbb{R}^3$ are derived.

Among these 15 contact relations, the relations that hold in convex cases are limited to those derived from the topological region-body relations that hold in convex cases—i.e., the relations in Figs. 7A,B,J,M,N (Section 4.2). At this time the relations in Figs. 7B,J,N are mapped to not three, but two contact relations, respectively, because the boundary of a convex body may contact either side of a convex region, but not both sides. Consequently, 2×1+3×2 = 8 contact relations, 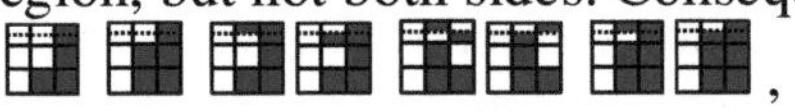, hold between a convex region and a convex body in $\mathbb{R}^3$.

5.3. Line-Region Relations

The mapping from topological line-region relations to contact relations is a bit complicated, because the line's interior and boundary may contact the region's either or both sides, in some cases not independently. To simplify, we start from the relations between a convex line and a convex region. Section 4.2 found that there are 13 topological line-region relations that hold in convex cases (Figs. 6a-c,f,j,l-p,s,C,I). Among these relations,

- the five relations in Figs. 6a,b,j,m,n, where the region's interior does not intersect with the line's interior or boundary, are mapped to one contact relation, respectively (and thus 5×1 contact relations   are derived);
- the seven relations in Figs. 6c,f,l,o,p,s,I, where the region's interior intersects with the line's interior/boundary, are mapped to two contact relations in which the line's interior/boundary contacts either side of the region (and thus 7×2 contact relations  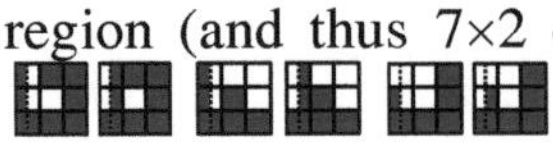 are derived); and

- the relation in Fig. 6C cannot be mapped to a contact relation because it requires piercing.

In this way, $5\times1+7\times2 = 19$ contact relations between a convex line and a convex body in $\mathbb{R}^3$ are derived.

If the line and the body are not limited to convex ones, contact relations are derived as follows:

- the seven topological relations in Figs. 6a,b,g,h,j,m,n, where the region's interior does not intersect with the line's interior or boundary, are mapped to one contact relation (e.g., ), respectively;
- the seven topological relations in Figs. 6c,i,k,o,B,C,H, where the region's interior intersects with a part of the line's interior and not the line's boundary, are mapped to three contact relations (e.g., ) in which the line's interior contacts the region's either or both sides, respectively;
- the two topological relations in Figs. 6D,E, where the region's interior intersects with both of the line's endpoints and not the line's interior, are mapped to three contact relations (e.g., 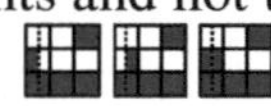) in which the line's two endpoints contact the region's either or both sides, respectively;
- the four topological relations in Figs. 6I-L, where the region's interior intersects with one of the line's endpoints and not the line's interior, are mapped to two contact relations (e.g., 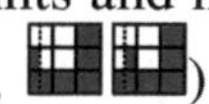) in which the line's one endpoint contacts the region's either side, respectively;
- the three topological relations in Figs. 6f,l,s, where the region's interior contains the line's interior, are mapped to two contact relations (e.g., 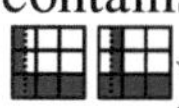) in which the line's interior contacts either side of the region and the line's endpoint(s) follows it, respectively;
- the two topological relations in Figs. 6d,A, where the region's interior intersects with both of the line's endpoints and a part of the line's interior, are mapped to 3×3 contact relations (e.g., 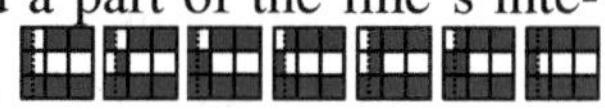 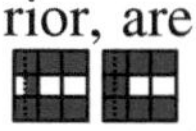) in which the line's interior and boundary contact either or both sides of the region independently, respectively;
- the four topological relations in Figs. 6p,q,F,G, where the region's interior intersects with one of the line's endpoints and a part of the line's interior, are mapped to 3×2 contact relations (e.g., 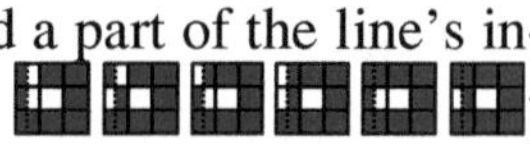) in which the line's interior contacts either or both sides of the region and the line's one endpoint contacts either side of the region independently, respectively;

- the topological relation in Fig. 6e is mapped to five contact relations (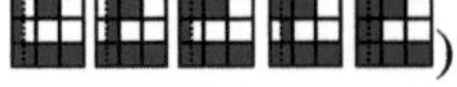) where the line's interior contact either or both sides of the region and the line's endpoints follow it; and
- similarly, the topological relation in Fig. 6r is mapped to four contact relations ().

In this way, in total 7×1+7×3+4×3+2×2+3×2+2×9+4×6+5+4 = 99 contact relations between a line and a region in $\mathbb{R}^3$ are derived

5.4. Region-Region Relations

The mapping from topological region-region relations to contact relations is more complicated than the previous case, because two sides and edge of one region may contact either or both sides of another region, in many cases not independently. The dependency occurs in the mapping from the topological relations in Figs. 8F-K,$P_{1/2}$-$W_{1/2}$ and those in Figs. 8d-$f_{1/2}$. In the former relations, the configurations always have enclosed space(s) (Section 4.4). Consequently, the side of one object that entirely faces the enclosed space cannot contact the side of another object that does not face the enclosed space. In the latter relations, one region *X* is contained by another region *Y*. Consequently, in their corresponding contact relations, if *X*'s either or both sides entirely contacts *Y*'s □-side, then *X*'s boundary also contacts *Y*'s □-side. Considering these constraints, we derived 1051 contact relations between two regions in $\mathbb{R}^3$ from the 43 topological region-region relations in $\mathbb{R}^3$ (See Appendix for the detail).

When two regions are limited to convex ones, the mapping becomes much simpler. Section 4.2 found 18 topological region-region relations that hold between two convex regions in $\mathbb{R}^3$ (Figs. 8a-$f_{1/2}$, B, E, and $L_{1/2}$-$O_{1/2}$). These relations are mapped to contact relations as follows:

- The two relations in Figs. 8a,b, which have no interior-related intersection, are mapped to one contact relation (e.g., 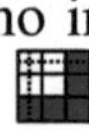), respectively;
- The six relations in Figs. 8c-$f_{1/2}$, which have an interior-interior intersection, are mapped to 2×2 contact relations (e.g., 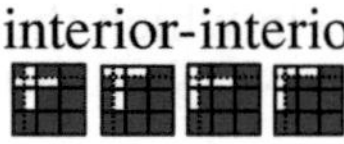) in which either side of one region contacts either side of another region, respectively;
- The four relations in Figs. 8$L_{1/2}$-$M_{1/2}$, which have a boundary-interior intersection but no interior-interior intersection, are mapped to two contact relation (e.g., 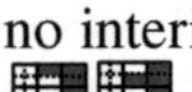) in which the boundary of one region contacts either side of another region, respectively;

- The six relations in Figs. 8B, E, and $N_{1/2}$-$O_{1/2}$ cannot be mapped to contact relation because they require piercing (Section 4.3).

Consequently, in total 2×1+6×4+4×2 = 34 contact relations between two convex regions in $\mathbb{R}^3$ are derived. These 34 contact relations are illustrated schematically in Fig. 12. This figure is actually a *conceptual neighborhood graph* (Freksa 1992) of the 34 contact relations, whose links show pairs of *similar* relations (in the sense that the configuration of one relation may switch to another relation by a minimal change). We can see nice horizontal and vertical symmetries in this graph, which result from the fact that each region has two sides.

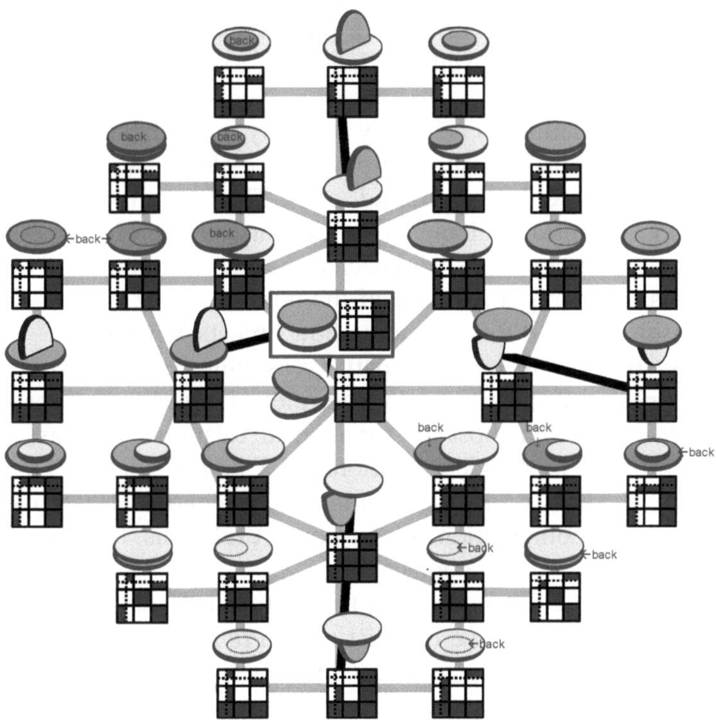

Fig. 12. A conceptual neighborhood graph of the 34 contact relations between two convex regions in $\mathbb{R}^3$

6. Conclusions and Future Work

This paper proposed a model of contact relations. The contact relations work as an alternative of topological relations when the target objects are solid. Due to the models' schematic similarity, contact relations have a strong correspondence with topological relations distinguished by the 9-intersection. Making use this correspondence, the sets of contact relations between various combinations of objects in $\mathbb{R}^3$ are derived from the corresponding sets of topological relations. The numbers of the derived relations (2 for point-body relations, 7 for line-body relations, 2 for body-body relations 15 for region-body relations, and 99 for line-region relations) are probably cognitively and practically acceptable, but unfortunately that for region-region relations (1051) is overwhelming. In general, if the objects are limited to convex ones, the numbers of the contact relations become much smaller (2 for point-body relations, 5 for line-body relations, 2 for body-body relations, 8 for region-body relations, 19 for line-region relations, and 34 region-region relations). One remaining issue is to analyze the completeness of the derived relations (i.e., to check the absence of additional relations).

This paper assumed that each body has only one surface element, but it is often possible to distinguish multiple surface elements of the body considering its shape (a cube, a pyramid, etc.). The 9^+-contact matrix allows the distinction of surface elements in arbitrary granularity, thanks to its nested structure. Of course, it is expected that the distinction of more surface elements yields rapid increase of the number of contact relations, but the model itself is useful for describing how objects contact.

One interesting future topic is spatial reasoning based on contact relations. It is expected that the composition rules of contact relations are considerably different from those of topological relations. For instance, the composition rule of topological relations "*if X includes whole Y and Y includes whole Z, then X includes whole Z*" (Egenhofer 1994) cannot be transformed to the composition rule of contact relations, because if *X* contacts whole *Y*, then *Y* cannot contact *Z*.

Another interesting future topic is to identify and analyze contact relations in $\mathbb{R}^2$. In a 2D space we can distinguish *left* and *right* sides of lines. 2D contact relations can be used for characterizing the spatial arrangements of two streets and, accordingly, reasoning techniques on such 2D contact relations will be useful for mobile robots to develop the knowledge of street networks efficiently without exploring all streets.

Acknowledgment

This work is supported by DFG (Deutsche Forschungsgemeinschaft) through the Collaborative Research Center SFB/TR 8 Spatial Cognition—Strategic Project "Spatial Calculi for Heterogeneous Objects."

References

Alexandroff, P.: Elementary Concepts of Topology. Dover Publications, Mineola, NY, USA (1961)

Billen, R., Kurata, Y.: Refining Topological Relations between Regions Considering Their Shapes. In: Cova, T., Miller, H., Beard, K., Frank, A., Goodchild, M. (eds.): GIScience 2008, Lecture Notes in Computer Science, vol. 5266, pp. 20-37. Springer, Berlin/Heidelberg, Germany (2008)

Billen, R., Zlatanova, S., Mathonet, P., Boniver, F.: The Dimensional Model: A Framework to Distinguish Spatial Relationships. In: Richardson, D., van Oosterom, P. (eds.): Advances in spatial data handling: 10th International Symposium on Spatial Data Handling, pp. 285-298. Springer, Berlin/Heidelberg, Germany (2002)

Egenhofer, M.: A Formal Definition of Binary Topological Relationships. In: Litwin, W., Schek, H.-J. (eds.): 3rd International Conference on Foundations of Data Organization and Algorithms, Lecture Notes in Computer Science, vol. 367, pp. 457-472. Springer, Berlin/Heidelberg, Germany (1989)

Egenhofer, M.: Deriving the Composition of Binary Topological Relations. Journal of Visual Languages and Computing 5(2), 133-149 (1994)

Egenhofer, M., Herring, J.: Categorizing Binary Topological Relationships between Regions, Lines and Points in Geographic Databases. In: Egenhofer, M., Herring, J., Smith, T., Park, K. (eds.): NCGIA Technical Reports 91-7. National Center for Geographic Information and Analysis, Santa Barbara, CA, USA (1991)

Feist, M.: On In and On: An Investigation into the Linguistic Encoding of Spatial Scenes. Ph.D. Thesis. Northwestern University (2000)

Freksa, C.: Temporal Reasoning Based on Semi-Intervals. Artificial Intelligence 54(1-2), 199-227 (1992)

Kurata, Y.: The 9^+-Intersection: A Universal Framework for Modeling Topological Relations. In: Cova, T., Miller, H., Beard, K., Frank, A., Goodchild, M. (eds.): GIScience 2008, Lecture Notes in Computer Science, vol. 5266, pp. 181-198. Springer, Berlin/Heidelberg, Germany (2008)

Kurata, Y., Egenhofer, M.: The 9^+-Intersection for Topological Relations between a Directed Line Segment and a Region. In: Gottfried, B. (ed.): 1st Workshop on Behavioral Monitoring and Interpretation, TZI-Bericht, vol. 42, pp. 62-76. Technogie-Zentrum Informatik, Universität Bremen, Germany (2007)

Mark, D., Egenhofer, M.: Modeling Spatial Relations between Lines and Regions: Combining Formal Mathematical Models and Human Subjects Testing. Cartography and Geographical Information Systems 21(3), 195-212 (1994)

Randell, D., Cui, Z., Cohn, A.: A Spatial Logic Based on Regions and Connection. In: Nebel, B., Rich, C., Swarout, W. (eds.): 3rd International Conference on Knowledge Representation and Reasoning, pp. 165-176. Morgan Kaufmann, San Francisco, CA, USA (1992)

Schneider, M., Behr, T.: Topological Relationships between Complex Spatial Objects. ACM Transactions on Database Systems 31(1), 39-81 (2006)

Shariff, A., Egenhofer, M., Mark, D.: Natural-Language Spatial Relations between Linear and Areal Objects: The Topology and Metric of English-Language Terms. International Journal of Geographical Information Science 12(3), 215-246 (1998)
Zlatanova, S.: On 3D Topological Relationships. In: Erwig, M., Schneider, M., Sellis, T. (eds.): International Workshop on Advanced Spatial Data Management, pp. 913-924. IEEE Computer Society (2000)

Appendix: Mapping from Topological Region-Region Relations in $\mathbb{R}^3$ to Contact Relations

- the topological relations in Figs. 8a,b,F are mapped to only one contact relation, respectively, due to the lack of interior-related intersections;
- the topological relations in Fig. 8c, Figs. 8A-B, Figs. 8C-D, Fig. 8E, Figs. 8$L_{1/2}$-$M_{1/2}$, and Figs. 8$N_{1/2}$-$O_{1/2}$ are mapped to 15×3×3, 15, 3×3, 15×3×3, 3, and 15×3 contact relations, respectively, because in these mappings we can consider edge-side contacts and side-side contacts independently; that is,
  - if the topological relations have a boundary-interior intersection, they are mapped to $2^2-1 = 3$ types of edge-side contacts (i.e., the boundary of one region contacts either or both sides of another region); and
  - if the topological relations have an interior-interior intersection, this intersection is mapped to $2^4-1 = 15$ types of side-side contacts (i.e., either or both sides one region contact either or both sides of another region);
- the topological relations in Fig. 8G is mapped to six contact relations—four are the contact relations where *A*'s one side contacts *B*'s one side (i.e., ) and two are the contact relations where *A*'s two sides contact *B*'s different sides (i.e., );
- the topological relation in Fig. 8H is mapped to four contact relations in which the boundary of each region contacts either side of another region (i.e., 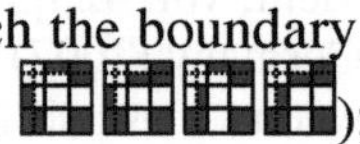);
- the topological relation in Fig. 8I is mapped to one more contact relation than the previous case, in which the boundary of each region contacts both sides of another region (i.e.,  in addition to );
- the topological relation in Fig. 8J is mapped to 24 contact relations, because once we assume that *A*'s *x*-side contacts *B*'s boundary and *B*'s *y*-side contacts *A*'s boundary (four possibilities exist), there are six possible side-side contacts (since *A*'s *x*-side and *B*'s *y*-side may contact either

or both sides of another region, *A*'s non-*x*-side may contact *B*'s *y*-side, and *B*'s non-*y*-side may contact *A*'s *x*-side);

- the topological relation in Fig. 8K is mapped to 15 more contact relations than the previous case, in which the boundary of both regions contact the both sides of another region and both sides of each region may contact either or both sides of another region;
- the topological relations in Figs. $8T_{1/2}$ are mapped to two contact relations, in which the entire boundary of one region contacts either side of another region;
- the topological relations in Figs. $8U_{1/2}$ are mapped to one more contact relation than the previous case, in which one region *clips* another region (Fig. 13a) and, accordingly, the boundary of the clipping region contacts both sides of the clipped region;
- the topological relations in Figs. $8P_{1/2}$ are mapped to four contact relations, in which the boundary of each region contacts either side of another region;
- the topological relations in Figs. $8Q_{1/2}$ are mapped to three more contact relations than the previous case, in which one region clips another region and the boundary of the clipped region contacts either or both sides of another region;
- the topological relations in Figs. $8R_1,V_1$ are mapped to 28 contact relations, because once we assume that *A*'s *x*-side and *B*'s *y*-side encloses a space (four possibilities exist), there are seven possible side-side contacts (because *A*'s *x*-side may contact *B*'s side, while A's non-*x*-side may contact either or both sides of *B*), and similarly the topological relations in Figs. $8R_2,V_2$ are mapped to 28 contact relations;
- the topological relations in Figs. $8W_{1/2}$ are mapped to 15 more relations than those in Figs. $8V_{1/2}$ are, in which either or both sides of the clipped region contact either or both sides of another regions;
- the topological relations in Figs. $8S_{1/2}$ are mapped to 45 more relations than those in Figs. $8R_{1/2}$ are, in which the boundary and either or both sides of the clipped region contact either or both sides of another region independently;
- the topological relation in Fig. 8d is mapped to four contact relations in which either side of one region contacts either side of another region;
- the topological relation in Fig. $8f_1$ is mapped to 16 contact relations, among which four are the relations in which one entire side of *A* contacts either side of *B* (i.e., ) and twelve are the relation where one side of *A* entirely contacts either side of *B* and *A*'s another side partly contacts either or both sides of *B* (i.e., 

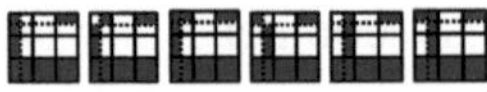), and similarly the topological relation in Fig. 8f2 is mapped to 16 contact relations; and

- the topological relation in Fig. 8e1 is mapped to eight more contact relations than that in Fig. 8f1, in which *A* is clipped by *B* plainly such that *B*'s one side contacts both sides of *A* (Fig. 13b) and *B*'s another side may contact either or both sides of *A*, and similarly, the topological relation in Fig. 8e2 is mapped to eight more contact relations than that in Fig. 8f2.

(a) (b)

Fig. 1. Configurations of two contact relations, derived from the topological region-region relations in (a) Fig. 8U_1 and (b) Fig. 8e_1, in which the region *A* is clipped by the region *B*

Needs and potential of 3D city information and sensor fusion technologies for vehicle positioning in urban environments

Marc-Oliver Löwner[1], Andreas Sasse[2] and Peter Hecker[2]

[1] Institut für Geodäsie und Photogrammetrie, Technische Universität Braunschweig. m-o.loewner@tu-bs.de
[2] Institut für Flugführung, Technische Universität Braunschweig. a.sasse@tu-bs.de, p.hecker@tu-bs.de

Abstract. We proposed a concept for an absolute vehicle positioning architecture in urban environments by accomplishing 3D city data and car side laser scanner. Today there is no reliable technology for a long time stable and absolute positioning in urban environments. While GNSS suffers from outtakes, SLAM and dead reckoning technologies are only precise during a limited time span. Because car mounted laser scanner produce bevel cuts, real 3D façade information is needed to apply a feature extraction algorithm. Our approach is based on a web feature service connecting the vehicle to a 3D city database. Starting from an initial position, a spatial request is phased to the database and geometries are sent to the car. There, feature extraction and map matching algorithm support an absolute positioning.

1. Introduction

While traffic situation in fast growing cities with increasing advent of individual traffic is going to become confusing, driver assistance systems play a more important role in safety concepts of carmakers. We will discuss the shortcomings of today's approaches and propose a solution for these. Our approach combines up-to-date 3D city data with feature extraction algorithms on the vehicle.

T. Neutens, P. De Maeyer (eds.), *Developments in 3D Geo-Information Sciences*,
Lecture Notes in Geoinformation and Cartography, DOI 10.1007/978-3-642-04791-6_8,

Today, advanced driver assistant systems (ADAS) are already able to support the driver in many situations like automatic distance control in stop-and-go scenarios, emergency braking or during lane changes [1]. All these systems have one thing in common: they take their decisions based on sensors which are observing the surrounding environment relative to the car. This approach is feasible as long as the ADAS is working in a non-collaborative way.

In the future, Car-2-Car communication technology will allow a wide range of new ADAS applications based on the exchange of detected obstacles or the current position of the own car [2]. The next step in the development of ADAS assumes that all positions are known in a global and therefore exchangeable reference frame. A fundamental precondition for a transformation of, for example, observed obstacles into a global representation is a well-known ego-state of the car. This ego-state usually includes the global position and velocity as well as the attitude.

Up to now, a global positioning was only necessary for the strategic routing process within the vehicle's navigation system. The use of the vehicle state for driver assistance systems, especially for safety critical ones, leads therefore to increasing requirements concerning the reliability of such information. These requirements can be divided into accuracy, availability, continuity and integrity of the determined states. In contrast to conventional navigation systems a highly accurate vehicle state has to be continuously available to the ADAS application. Furthermore, the integrity of the provided information plays a much more important role due to the fact that this application will not only provide information but warnings or, to some extend, automatically initiates a control input to the car (e.g. emergency braking).

Today digital information about infrastructure is available and could support the absolute positioning of a vehicle under the assumption of the use of additional sensors. The main source in Germany is the Automatisierte Liegenschaftskarte (ALK) which was initiated by the Arbeitsgemeinschaft der Vermessungsverwaltungen der Länder der Bundesrepublik Deutschland (AdV) in the early 80ies. Next to properties, land parcels, and streets it covers detailed geometries about and street furniture like trees and gully covers etc. Geometries are associated with a catalogue of objects representing a semantically data model. The ALK is a 2D-based digital information system that incompletely holds information about building heights. Therefore the ALK can serve as one data input for a vehicle positioning system though it is insufficient in terms of 3D information.

A true 3D-data model for virtual cities was developed by the Special Interest Group 3D (SIG 3D) of the initiative Geodata Infrastructure North-Rhine Westphalia, Germany and first presented by [3] in German. Today it has become a standard of the Open Geospatial Consortium (OGC) [4] and is well documented in [5]. Within well defined level-of-details from LoD0 to LoD4 with increasing accuracy and structural information, CityGML presents an information model to describe the semantics and geometry of virtual 3D city models. While LOD1 as a blocks model can be roughly produced by the use of 2D footprints and building height, sophisticated data acquisitioning procedures are needed to describe more detailed information about building façades stored in LoD3.

This paper describes needs and potentials for an absolute vehicle positioning in urban environments. We follow an approach that combines 3D infrastructural information with sensor fusion technologies on the vehicle. Therefore we first review the problems of absolute positioning techniques in urban environments without additional city data. Then we argue the problems of 2D data for this approach (Chap. 2). In chapter 0 we propose an overall concept for a positioning approach. This includes data integration from different sources, data transfer via a Web Feature Service and information fusion on the vehicle.

2. Problems and Limitations for an absolute vehicle positioning

This section focuses on problems and limitations of current and future localisation technologies. First, a short review of state-of-the-art positioning systems is presented, followed by a discussion of major shortcomings.

2.1. State-of-the-art of localization technologies

Global vehicle positioning is usually done by means of satellite navigation. The most popular global navigation satellite system (GNSS) is the Global Positioning System (GPS) operated by the U.S. department of defence. Applying this system the user is able to localize itself with an accuracy of 10 meters depending on the environmental conditions [6]. Especially in urban environments the occurrence of disturbances can increase dramatically. **Fig. 1** summarizes the most important error sources.

To overcome drawbacks of GNSS, complementary sensor like odometers, steering encoders and in some automotive applications inertial measurement units (IMU) are integrated with satellite navigation. This usually requires that the entire localization system is part of the vehicle's standard equipment. The basic principle applied by this group of sensors is known as dead reckoning (DR) describing the integration of measurement values (e.g. velocity, acceleration) over time to get a position estimate [7]. In a combined mode, the position accuracy compared to stand-alone GNSS is enhanced, but the exclusive use of DR sensors would lead to a position drift depending on the sensor quality.

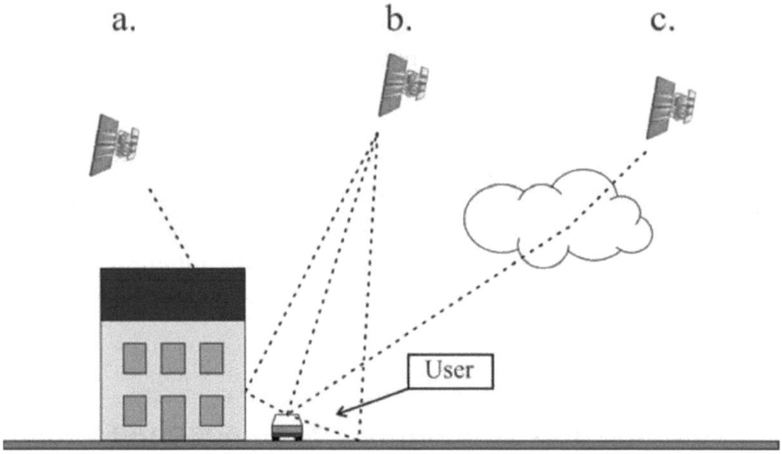

Fig. 1: Most important error sources of GNSS, namely a) shading, b) multi path effects and c) atmospheric effects.

In addition to this conventional approach of fusing GNSS with DR sensors a third group of localization systems is especially known to the robotic community. There, visual sensors like laser scanners or cameras are used to extract landmarks (e.g. walls or corners) from a single scan or frame. These landmarks are tracked over time and used to re-localize the user as well as the landmark itself. This approach leads to a simultaneous localization and mapping (SLAM) of the user and the landmarks respectively [8], but requires a continuous tracking of landmarks.

2.2. Major shortcomings in urban environments

SLAM benefits from the fact that non a-priori knowledge like landmark positions is necessary. On the other hand this fact limits the approach to relative positioning. The use of additional information like detailed maps can lead to an absolute positioning system as illustrated in Fig. 2. In this test arrangement a laser scanner was mounted horizontally on the roof of a test vehicle which turns the localization problem into a 2D one. A high precision digital city map provides geographically referenced positions of buildings and even trees. By matching this a-priori knowledge with the observed landmarks, absolute positioning becomes possible. However, this procedure needs to assume that the extrusion of building footprints into 3D is a valid one.

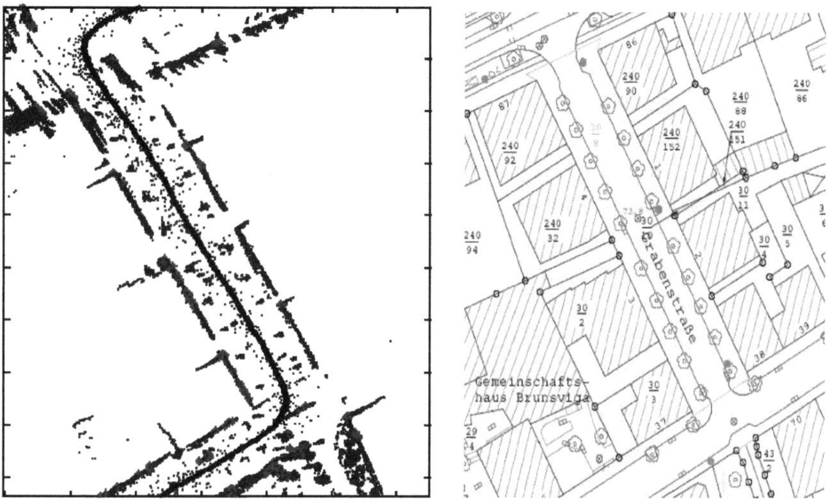

Fig. 2: Example of a 2D localization system using a digital city map in combination with a horizontal 2D laser scanner. The left figure shows the accumulated raw data, the right one a cut out from the digital city map.

In practice, this 2D approach suffers from the fact that laser scanners are usually situated in the bumper of a car. Due to the horizontal alignment of the laser scanner the resulting scan images in an urban environment like a residential area would be mainly dominated by parked cars and other non-infrastructural obstacles. The observed geometries are therefore not included in any map due to the fact that they are non-permanent objects. This problem can be solved by rotating the laser scanner plane as described in more detail in chapter 0. In this case the a-priori knowledge needs to be extended to 3D.

Table 1: Sensor complementarities

	SLAM	GNSS	DR	Map/Lidar
drift	yes (-)	no (+)	yes (-)	no (+)
geo. reference	no (-)	yes (+)	no (-)	yes (+)
ext. disturbance	no (+)	yes (-)	no (+)	no (+)
data rate	medium (+/-)	low (-)	high (+)	medium (+/-)
urban performance	high (+)	low (-)	high (+)	high (+)
free field performance	low (-)	high (+)	high (+)	low (-)

Basically the resulting 3D city map / laser scanner localization system has additional complementary properties to conventional GNSS / DR integrations. An overview can be found in Table 1. A robust positioning system for safety critical ADAS expects at least two positive qualities per row to keep redundancy in the sense of fault detection.

Especially in an urban environment, where GNSS outages can occur frequently, the use of laser scanner measurements in combination with appropriated a-priori knowledge can enhance the overall system accuracy and, more important to ADAS applications, availability. Furthermore integrity algorithms will profit from redundant position estimates in the case of GNSS and city map / laser scanner availability.

2.3. Need for up-to-date 3D data

3D building representation can be roughly produced by the fusion of building footprints and height data. Whereas building footprints are stored in ground plans like the ALK, height information can be obtained from floor numbers or airborne laser scanner data. This is a current procedure when producing CityGML's LoD1 building representation (e.g. [9]). A major problem is the incorporation of 3D data, especially information about the façades. Even the building's vertical bounding edges are hardly to detect using 2D ground plans. In the first approach outlined in fig. 0.2, horizontal information is extrapolated to 3D space withoutin information about building heights and other error sources. When using the ALK, only under roof surfaces can be identified. There is no direct information about heights of gangways, which leads to errors in assuming information about the façade. **Fig. 3** illustrates a typical error case when extruding the ALK to the third dimension. While the 3D model shows the geometry of a gangway, it is only betoken in the 2D dataset of the ALK. These errors would lead to misinterpretations of the laser supported positioning system of the vehicle.

Extruding 2D data to 3D façade representation definitely fails when detailed information is required. While a vehicle launched laser scanners and according edge extraction algorithms are able to detect ridges in a feasible resolution, window apertures and doorways are of interest. These features can be stored in CityGML's LoD3 building model. The incorporation of as many vertical and horizontal edges as possible will enhance plausibility of the positioning algorithm.

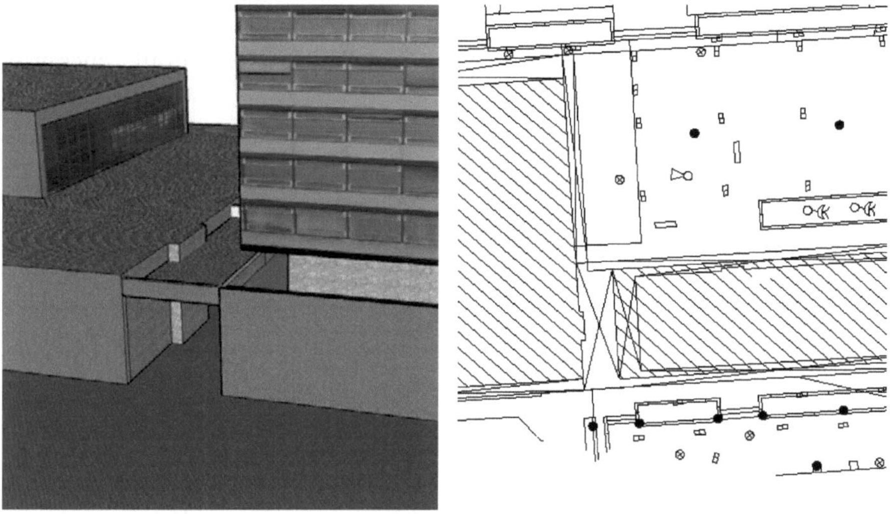

Fig. 3: Gangway represented left in a true 3D model and right in the ALK. It is only implied by the ALK but no height information is available.

Next to precision, actuality of 3D data is essential for a reliable positioning solution. In fast growing and changing urban environments, up to 10 % of infrastructure changes within one year. Change rate of 3D city information often changes even faster then 2D data [10]. While this statement is made for roof surface geometry it is not clear how change rates are valid for building façades. Nevertheless, when incorporating footpaths, trees, traffic signs and other city furniture it is apparent that it is deficient to carry a static database on the vehicle. In this approach we suggest the development of a Web Feature Service to organize the data transfer from an integrated database to the car.

Overall concept for a service oriented navigation architecture

The above discussed limitations of common technologies for vehicle positioning in urban environments have to be overcome. We suggest a highly precise concept that links up vehicle sensors and 3D city data. It is meant to fulfil the requirements of advanced driver assistant systems relevant to safety and is complementary to global navigation satellite systems. Communication between the vehicle and an integrated database will be realized using a modified web feature service (WFS). The overall concept is depicted in **Fig. 4**.

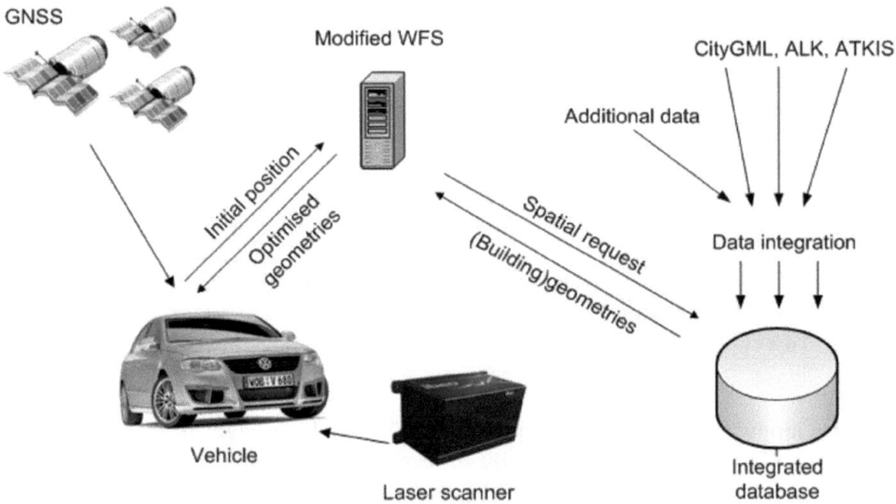

Fig. 4: Overall vehicle positioning concept.

A GNSS based initial position is transmitted from the vehicle to the WFS. The service phrases a spatial request that is communicated to a database holding information about the features relevant for the positioning algorithm. This database is a result of integration process incorporating typical data sources known for urban features. In our case this would be CityGML data if available, the German ALK, and additional data of façade information if needed. The geometries are sent back to the vehicle via the WFS. There, a map matching algorithm is applied to identify the position of the vehicle using the geometries identified by the on board laser scanner and the data as a-priori information. In a next step more information than the initial position can be transferred to the WFS causing different spatial requests on the database.

In the following we will discuss the concept and its problems in more detail. We will focus on first, the data integration process needed to allocate data relevant to the positioning concept. Second, we describe concepts for a vehicle-to-database service and third, we discuss the object extraction and information fusion on the vehicle. Our approach is meant to be an absolute and long-time stable positioning system.

3.1. The data integration problem

Today, data integration of city data from different sources still represents a major challenge. Reasons for this are heterogeneous database schemas as well as the middleware adapted to these schemas. In Germany many data sources for infrastructural information are available. Information about a building's footprint, footpaths, street furniture, and more are stored in the Automatische Liegenschaftskarte (ALK) or the Amtliches Topographisch-Kartographisches Informationssystem (ATKIS) [11]. While the latter is only available in parts of Germany today, both databases store absolute 2D coordinates. Furthermore, 3D city models of various qualities are under construction by order of public communities and GIS companies.

Evaluation of existing data sources and their ability to support the vehicle positioning approach is the first step that needs to be performed. Quality in terms of precision, actuality, and completeness has to be in the focus of recognition. As a result a catalogue of requirements can be compiled defining essential properties of data sets intended to use for the vehicle positioning approach.

None of the data schemas mentioned above may be used for positioning directly. While the German cadastre allocating the ALK and ATKIS is still 2D these data would need to be upgraded to the third dimension. Extruding a 2D footprint of a building, for instance, but raises the problems discussed in section 0. The lack of façade data is a not acceptable limitation for this approach. CityGML data on the other hand seams to be too complex in terms of semantics and geometry. As CityGML's LoD4 model of course would offer much information about façades, semantics and geometry of interior and roof structures are redundant. That applies to CityGML's appearance concept, also and is partly true for roof surface information encoded in CityGML's LoD3. Besides, the GML inherent overhead would causes needless data transfer costs. Although only very few data concepts are considered here, it seams to be clear, that additional data has to be acquired. It is clear that no sufficient information about of façades is available today.

The next step will be the development of a database schema. It has to meet the demands of an effective allocation of infrastructural data needed for vehicle positioning as façade information, street furniture, trees and so forth. This schema then is a starting point for data integration from known data sources. It is desirable to develop mapping tools for automatic integration of known sources. This would enable a nearly real time integration and update of the data basis for vehicle navigation. It is reasonable to consider a multi-scale approach. If so, the database directly supports an effective data transfer to the vehicle against demanded data precision, service agreement or available bandwidth for transmission.

3.2. A vehicle-to-database-service

Due to permanent changes of infrastructural data the database cannot be stored on the vehicle directly, but has to be accessed in real time. Therefore a vehicle-to-database-service that is based on an OGC Web Feature Service [12] has to be developed. This service organizes the bidirectional communication of the vehicle and the database. Further it has to interpret the request of the vehicle and organize the spatial request to the database. Finally the service will prepare the transmission of optimized infrastructure data to the vehicle. The transmission can be performed using the Universal Mobile Telecommunications System (UMTS).

The implementation specification of the vehicle-to-database-service needs to be capable to interpret different information types coming from the vehicle. In the initial state of a positioning procedure this will be an imprecise position which might be derived from GNSS measurements. As no additional information is available at this time, a bounding box algorithm will be applied to request the geometries needed. A time step further, the covered route and velocity may need to be converted. For the amount of data this would mean that a more precise spatial request can be applied. As the covered route is known, only half of the requested bounding box would apply for transmission. As many cars carrying traditional navigation systems the planed route can be transmitted via the well-known Geographic Data Files (GDF) [13]. Again this would mean to query a more restricted subset of data for the positioning procedure. However, as GDF data is only a planned route, not only the geometries of the streets transmitted have to be queried but the side roads as well. To conclude, the longer the positioning service runs, the more precise the spatial request can be performed. Therefore it is clear that the first request of the vehicle causes the most amounts of data that has to be transferred. With each step the transmission costs for spatial data required by the vehicle is going to decrease.

3.3. Object extraction and information fusion

Within the positioning procedure, a conventional system architectures consisting of a GNSS / DR fusion is combined with the city map / laser scanner approach. Hence, the complementary properties of both global positioning techniques mentioned above are utilized. It is expected that this combination leads to an effective enhancement of the localization system reliability especially in urban environments.

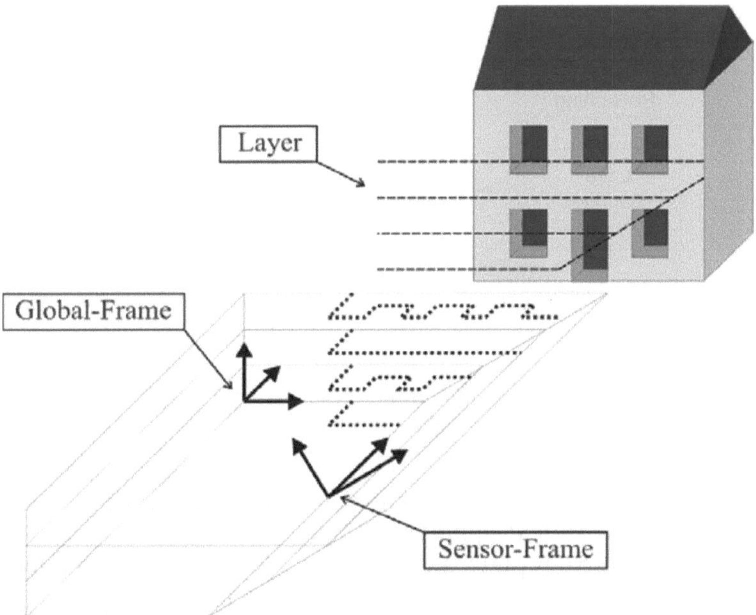

Fig. 5: 3D data acquisition using a single 2D laser scanner.

In a first step, the laser scanner's field of view problem mentioned before has to be solved. Therefore the measurement plane of a 2D laser scanner mounted at the front bumper of a car is screwed like illustrated in **Fig. 5**. The spatial allocation of the measurement points to horizontal section planes leads to a 3D image of the surrounding objects. Due to the rotation of the scan plane, the influence of non-permanent obstacles is substantially reduced. The resulting requirements for the laser scanner are close to today's automotive devices. Only the update rate has to be slightly higher.

After a batch mode collection of data, the measurement points are segmented within each section plane and between them. In a next step object edges are extracted in 3D from the resulting cloud of measurement points. Based on these edges a matching process with a 3D city model can be conducted. This leads to a relative position information with respect to an absolute referenced object. Finally, this information can be fused with GNSS and DR sensors. The resulting system architecture is presented in **Fig. 6**.

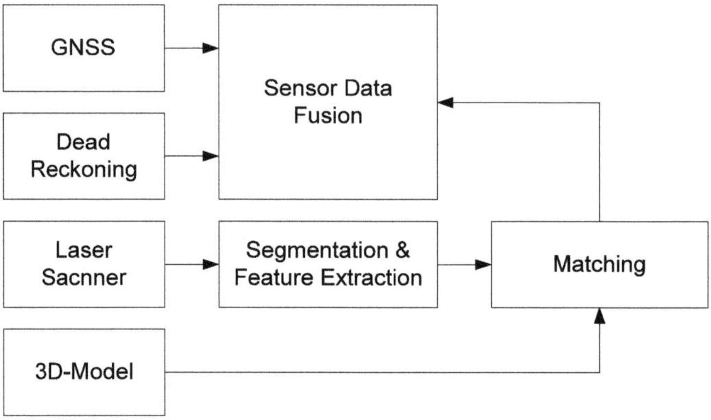

Fig. 6: Localization system architecture

Against the background of a potential use of such an urban localization system for safety critical applications like future ADAS, the integrity of the presented approach plays an important role. Therefore an appropriated vehicle autonomous integrity monitoring has to be implemented as part of the fusion algorithm.

Conclusion

We proposed a concept for an absolute and long time stable vehicle positioning architecture in urban environments. It benefits from accomplishing 3D city data and car side laser scanner and feature extraction algorithms. Data is requested directly from the vehicle and supported by a web feature service that communicates with an updated database holding integrated 3D city data from different sources.

We outlined the present state of positioning technology for vehicles in urban environment that is not able to support an absolute, long time stable and reliable location today. While global navigation satellite systems (GNSS) deliver sufficient positioning solutions in a free field situation, due to outtakes they are inadequate within an urban environment. These outtakes emerge from shadowing effects and multi path effects. Simultaneous localization and mapping (SLAM) and dead reckoning technologies are only applicable to the disadvantage of long time stability and an absolute geo referenced position.

We see the largest potential for absolute and long time stable vehicle positioning in the combination of map data and feature extraction algorithms. First results show that a 2D localization system using a digital city map in combination with a horizontal 2D laser scanner has the ability to support an absolute positioning architecture. Installation location of car mounted laser scanner and obstacles unfortunately necessitate the inspection of façade information by rotating the laser scanner plane when applying the feature extracting algorithm. Acquisition of a-priori data input for the positioning algorithm however cannot be done by just extruding 2D maps to the third dimension. Errors may happen while few characteristics like gangways are not marked in 2D maps sufficiently. Hence, façade information has to make assessable to the vehicle's feature extraction algorithm.

Our approach is meant to improve the discussed shortcomings for vehicle positioning in urban environments. It is based on a web feature service (WFS) that interconnects the vehicle with a 3D city database. Receiving a initial GNSS based position the service processes a spatial request sent back to the car. This approach assures the transmission of up-to-date data as it is stored on the data base. Data acquisition still is a major challenge. Although it is planed to integrate the most promising data formats like the German Automatisierte Liegenschaftskarte (ALK) or CityGML, it is likely, that additional data of building façades needs to be collected. Today, no sufficient information about façades are available that could support our approach directely.

The proposed service may be applicable as a real core for a Car-2-Car infrastructure. Once a communication infrastructure for vehicle and database is established, the car may serve for data supplier for temporal data like density of traffic or working sites itself. It can easily recognized by subtracting the effective laser scanner data from the 3D features delivered from the integrated data base. This information itself can be stored on the database and referred to other vehicles. Besides, information of changes in city geometry would assist data providing companies to keep their datasets in a current state.

References

[1] Winner, H., Hakuli, S., Wolf, G., 2009: Handbuch Fahrerassistenzsysteme, Vieweg+Teubner Verlag, Wiesbaden.

[2] Röckl, M., Gacnik, J., Schomerus, J., 2008: *Integration of Car-2-Car Communication as a Virtual Sensor in Automotive Sensor Fusion for Advanced Driver Assistance Systems.* Proceedings, Springer Automotive Media, FISITA 2008, Munich, Germany.

[3] Gröger, G., Benner, J., Dörschlag, D., Drees, R., Gruber, U., Leinemann, K. and Löwner, M.-O., 2005: Das interoperable 3D Stadtmodell of der SIG 3D. Zeitschrift für Geodäsie, Geoinformation und Landmanagement, 6/2005: 343-353.

[4] Gröger, G.; Kolbe, T. H.; A. Czerwinski and C. Nagel 2008: OpenGIS City Geography Markup Language (CityGML) Implementation Specification.

[5] Kolbe, T. H., 2008: Representing and exchanging 3D City Models with CityGML. In: Lee, J. and Zlatanova S. (Eds.): 3D Geo-Information Sciences. Berlin, p 15-31.

[6] Parkinson, B. W., Spilker, J.J. Jr., 1996: *Global Positioning System: Theory and Applications*, vols. 1 and 2, American Institute of Aeronautics, Washington, DC.

[7] Titterton, D. H., Weston, J. L., 2004: *Strapdown Inertial Navigation Technology*, vol. 207, American Institute of Aeronautics, Washington, DC.

[8] Thrun, S., Burgard, W., Fox, D., 2005: *Probabilistic Robotics*, MIT Press.

[9] Schwalbe, E., Maas, H.-G., Seidel, F. 2005: 3D building model generation from airborne laser scanner data using 2D GIS data and orthogonal point cloud projections. ISPRS Workshop on "Laser scanning 2005", Netherlands.

[10] Brenner, C. (2005): Building reconstruction from images and laser scanning - In Int. Journal of Applied Earth Observation and Geoinformation, Vol. 6(3-4): 187-198.

[11] AdV, 2008: Dokumentation zur Modellierung der Geoinformationen des amtlichen Vermessungswesens (GeoInfoDok). 357 pp. http://www.adv-online.de/extdeu/broker.jsp (Accessed May 2009)

[12] Vretanos, P., 2004: OpenGIS Web Feature Service Implementation Specification Version 1.1.0. OpenGIS Project Document 04-094 http://www.opengis.org (Accessed May 2008)

[13] International Organization for Standardization 2004: Intelligent transport systems - Geographic Data Files (GDF) - Overall data specification.

Modeling Visibility through Visual Landmarks in 3D Navigation using Geo-DBMS

Ivin Amri Musliman, Behnam Alizadehashrafi, Tet-Khuan Chen and Alias Abdul-Rahman

Dept. of Geoinformatics, Faculty of Geoinformation Science & Engineering, Universiti Teknologi Malaysia, 81310 Skudai, Johor, Malaysia.
ivinamri@utm.my, behnam5us@gmail.com, kenchen14@gmail.com, alias@utm.my

Abstract. Today's map navigation systems (from 2D to 3D) provide direction instructions in the form of maps, pictograms, and spoken language. However, they are so far not able to support or has very limited access to landmark-based navigation, which the most natural navigation concept is for humans and which also plays an important role for upcoming personal navigation systems. In order to provide such navigation, in this paper, we discuss one of possible solution of modeling visibility in 3D navigation through visual landmarks using Geo-DBMS approach. The aim is to generate measurable visual landmarks along the focus map in a city model which can be used in car or pedestrian navigation system (as web or mobile application). The focus map is obtained from 3D analytical operation (3D buffering from the 3D shortest path analysis result) function within Geo-DBMS. Detailing to the generated measurable visual landmark's façade, an implementation of dynamic pulse function is then applied. The techniques for choosing specific landmarks and generating the focus maps are shortly presented and their functionality is explained. We tested the proposed approach by using Stuttgart 3D city model. Finally, the paper provides outlook on ideas for future deployment and research.

T. Neutens, P. De Maeyer (eds.), *Developments in 3D Geo-Information Sciences*,
Lecture Notes in Geoinformation and Cartography, DOI 10.1007/978-3-642-04791-6_9,

Key words. Visual landmarks, focus map, dynamic pulse functions, 3D navigation, 3D buffer.

1. Introduction

Enhanced 2D and 3D maps, landmarks and focus based maps play an important role within map routing in navigation applications especially in route finding. Landmarks were originally defined as clearly visible geographic features that help people to orientate themselves in the field. In modern usage, the term landmark has been extended to anything that is easily recognizable, such as a building, famous historical objects, monument or a distinctive large tree. Currently, visual landmarks are better known as Points of Interest (POI) in particular to map navigation and web mapping applications (e.g. car navigation systems, personal navigation devices, digital travel and city guides, etc.), but not as landmark-based navigation which were hardly supported and accessed in any map navigation systems. It is known that landmarks have to be associated in local maps, a local map being a set of landmarks put together because of (1) memory constraints arising from the use of embedded mobile systems, (2) the need to download updated information from a server, and (3) essential to compute a location based on landmark orientation.

This paper addresses the question of enriching visibility in navigation using visual landmarks within Geo-DBMS environment. Although there are no specific guidelines regarding how to model visibility through visual landmarks, past relevant researches are more towards image processing for computer vision and robotics as described by Langdon et. al. (2006) and Se et. al. (2001). Others are focusing on information in the area of user is currently interested in - e.g. through the concept of "Focus Maps" (Zipf and Richter 2002) and adding landmarks as key elements of way finding support (proposed by Golledge 1999). The concept of focus maps (Zipf & Richter 2002) briefly indicates that some regions on the maps are usually having higher interest to the user and should therefore presented in a more dominant way especially in the case of maps on mobile devices with limited displays. In 2007, Neis and Zipf proposed a focus route map with landmarks which can be obtained by generating two buffers using DWITHIN function in WFS filter in Open Location Service (OpenLS) environment. The inner buffer is responsible for detection of the landmarks and focusing the map, while the second buffer is supporting the focus algorithm.

Therefore the following question arises: How can we realize visual landmarks and focus-based maps in 3D navigation system? Since the targeted environment in 3D, what is the best tool can be used to store, retrieve and manipulate these visual landmarks but within the context of wayfinding?

To adapt visual landmarks into map navigation, evaluations are needed to rate the importance of the integrated map objects. Important or distinguishable landmarks can be visualized more prominent than other less relevant information. This differentiation can be apparent by undergo generalization process or assigning different coloring of objects. Zipf (2002) emphasizes the inclusion of situational parameters like personal preferences or general context parameters.

The paper is organized in the following order: first, short discussion of the current visual landmark direction in wayfinding. Section 2 gives an overview of navigation using landmarks and highlights the importance of landmarks for navigation. In section 3 we present the properties of features used to measure their attractiveness as visual landmarks and introduction to dynamic pulse function. Then, the implementation of geo-DBMS approach for the 3D analytical analysis in modeling visibility of visual landmarks will be discussed in Section 4. The solution involving the method to create new 3D buffering function is discussed in this section. The experiment and discussions are presented in Section 5 and the research is concluded with some future work remarks in Section 6.

2. Navigation Using Landmarks

Modern car navigation systems have been introduced in 1995 in upper class cars and are now available for practically any model. They are relatively complex and mature systems able to provide route guidance in form of digital maps, driving direction pictograms, and spoken language driving instructions (Zhao, 1997). Obviously, the introduction of buildings as landmarks together with corresponding spoken instructions (such as 'turn right after the tower') would be a step towards a more natural navigation.

Thus, the main problem lies in identifying suitable landmarks and evaluating their usefulness for navigation instructions.

There are two different kinds of route directions to convey the navigational information to the user: either in terms of a description (verbal instructions) or by means of a depiction (route map). According to (Tversky and Lee, 1999) the structure and semantic content of both is equal, they consist of landmarks, orientation and actions. Using landmarks is important, because they serve multiple purposes in way finding: they help to organize space, because they are reference points in the environment and they support the navigation by identifying choice points, where a navigational decision has to be made (Golledge, 1999). Accordingly, the term landmark stands for a salient object in the environment that aids the user in navigating and understanding the space (Sorrows and Hirtle, 1999). In general, an indicator of landmarks can be particular visual characteristic, unique purpose or meaning, or central or prominent location.

Furthermore, landmarks can be divided into three categories: visual, cognitive and structural landmarks. The more of these categories apply for the particular object, the more it qualifies as a landmark (Sorrows and Hirtle, 1999). This concept is used by (Raubal and Winter, 2002) to provide measures to specify formally the landmark saliency of buildings: the strength or attractiveness of landmarks is determined by the components visual attraction (e.g. consisting of facade area, shape, color, visibility), semantic attraction (cultural and historical importance, explicit marks, e.g. shop signs) and structural attraction. The combination of the property values leads to a numerical estimation of the landmark's saliency.

Landmark saliency of a feature does not depend on its individual attributes but on the distinction to attributes of close features. Being a landmark is a relative property. Landmarks are used in representations of space and in the communication of route directions (Werner et. al. 1997). Route directions shall provide a set of procedures and descriptions that allow someone to travel. Studies show that landmarks are selected for route directions preferably at decision points (Michon and Denis, 2001). Another study has shown that mapped routes enriched with landmarks at decision points lead to better guidance, or less wayfinding errors, than routes without landmarks. Furthermore, different methods of landmark presentations were equally effective (Deakin 1996). Lynch (1960) defines landmarks as external points of reference points that are not part of a route like the nodes in a travel network.

Sorrows and Hirtle (1999) categorize landmarks into visual (visual contrast), structural (prominent location), and cognitive (use, meaning) ones, depending on their dominant individual quality. A landmark will be stronger the more qualities it possesses. However, a formal measure for the landmark saliency of an object is still missing.

3. THE METHODOLOGY

3.1. Measures for the Attractiveness of Landmarks

The main contribution of this paper is a formal model of landmark saliency, which includes measures for the attractiveness of landmarks. In order to determine whether a feature qualifies as an attractive landmark we specify properties, which determine the strength of a landmark. In this section we identify such measures by taking into account the three types of landmarks as proposed by Sorrows and Hirtle (1999). Following their framework we are only focusing on the modified visual attraction of features in geographic space to determine their use as visual landmarks.

3.1.1. Visual Attraction

Landmarks qualify as visually attractive if they have certain visual characteristics such as a sharp contrast with their surroundings or a prominent spatial location. Our formal model of landmark saliency includes two measures criteria regarding visual attraction: shape volume and shape complexity.

3.1.1.1. Shape Volume

The shape volume of an object is an important property for determining its contrast to surrounding objects. People tend to easily notice objects whose shape areas significantly exceed or fall below the shape areas of surrounding objects such as largest volume of buildings or tallest building. In the trivial case of a regularly shaped building (of rectangular form) the area is calculated by multiplying its width and height.

In Geo-DBMS, the volume can be obtained by using this SQL function:

```
SELECT Calc3DVolume (Face1, Face2, Face3, Facen) FROM
   Building3DSolid_Table;
```

3.1.1.2. Shape Complexity

Visual attraction of an object is also determined by its complexity of shape. Unorthodox shapes amidst conventional rectangle-like shapes strike one's eyes. We formally specify the shape complexity measure of an object by considering its shape factor and also the deviation of its shape from that of a rectangle. The shape complexity factor represents the proportion of height and width and the 3D data type being used to model it in Geo-DBMS. Complex shapes of a building will result in combination of different types of 3D object data types. For example, skyscraper with a huge parabolic structure at the top will have a higher shape factor and required a sphere and multipolygon data type, whereas long and low buildings will have a low shape factor and use simpler multipolygon data type structure.

In Geo-DBMS, the shape complexity can be determined using this SQL function:

```
SELECT validate_SOLID3D(GEOMETRY) VALID from test where
PID>1;
```

The result:

```
PID   VALIDATION_SOLID3D
----------------------
 1     TRUE
 2     TRUE
 3     TRUE
```

From the result above, it is shown that this particular multipolygon data type combines three solid objects which represent a building or a landmark. To determine which landmark has most complex structure is based on the sum of numbers of returned true validation of 3D solid objects. Then it will be sorted again by calculating its 3D volume.

3.2. Dynamic Pulse Function

One of ongoing research is the MoNa3D which implementing synthetic texturing and dynamic pulse functions for 3D navigation purpose targeted for mobile platform (Coors and Zipf, 2007). This function is 'similar' to the Levels of Detail (LoD) approach. In the implementation of LoD 2 and 3 in VRML, details of the façade of an object (building) can be textured by adding images and 'draped' on a particular façade surface. Therefore, loading time in web browser of the final model will depends on the size of the VRML file (*.wrl) and the size of images that are being used to textured each façade. The bigger the file size of each entity, the longer it takes to load in a web browser. In this paper, other possible approach is implemented, e.g. the dynamic pulse function that will resulting in low file size of a final model of a fully textured object (building) which is similar to LoD 2 or 3 quality.

The details of this function are described as follows. Each building consists of polygons and for each polygon there are some pulse functions in x and y axes which controls the process of texture generation. Pulse function can be easily defined with the logical values. In intersection of X and Y axes, TRUE means insert and FALSE means not to insert the texture (Coors and Zipf, 2007). In order to increase the flexibility, the number of layers of textures and pulse functions are unlimited and each of them has its own priority in respect to others. In the process of creating database for synthetic texturing method, we have to define lots of fields like number of facades which are generating an identical building, texture file name and path or directory; pulse functions and their related parameters and so on. According to our algorithm the group of windows which are exactly in the same rows, columns, size, shape and vertical or horizontal distances can be assumed as one layer. By observing each facade and its rectified image we can make a decision on number of layers for the windows.

For example in figure 1: we have five layers according to the rules that we already mentioned for the windows. Layer 3 and layer 4 cannot be merged to one layer because they don't have exactly the same horizontal cluster or distance even if they have the same shape and size and vertical distance. Now it's time to generate the dynamic pulse function or the size of output image file and frame in respect to the size of rectified image. After generating the frame of output image file using a java program, we have to measure the geometries of components in respect to the upper left corner of the rectified picture and entered the parameters inside the JavaScript program and finally running it in web browser. According to the number of window layers that we observed plus one layer for the wall, we do need to run the JavaScript program 5 times. Indeed, we will have five XML schema files with repeated wall layer and if we make union of all of these files, the XML result schema concludes 6 layers.

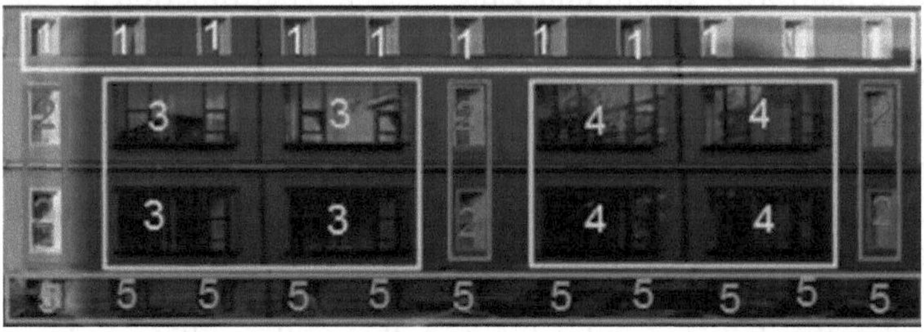

Figure 1: By observation on the rectified facade we can decide the number of layers. In this example we need 5 layers for the windows and one layer for the wall.

In our application which is programmed in JavaScript and can be run in web browser, we can input parameters about the geometries of components in respect to coordinate of the rectified façade image from upper left corner along with the texture file names extracted from perpendicular images from windows and doors. If we run this program, it will generate the parameters in respect to the dynamic pulse function and save them in a database as a XML schema and then the XML schema file can be run with our java program to generate the façade semi automatically. Reasonable deference can be seen in the size and quality of rectified image and our output image file. The size of the picture taken by 10 megapixel wide angel digital camera is (3888 * 2592) 2.85MB and the rectified image size is (8348 * 2854) 6.93MB with lots of leaning geometries and lack of quality in farther area from exposure point and disturbing objects like pedestrians and parking cars. In the other hand the size of our output image file is just (747 * 256) 35.4KB without disturbing obstacles and with a very high quality.

Figure 2: From left to right wall texture (30*256) 25.9KB, window texture for layer 1 (251*273)35.3KB, window texture for layer 2 (282*556)49.4KB, window texture for layers 3 and 4 (1053*586)83.6KB, window texture for layer 5 (481*330)45.4KB

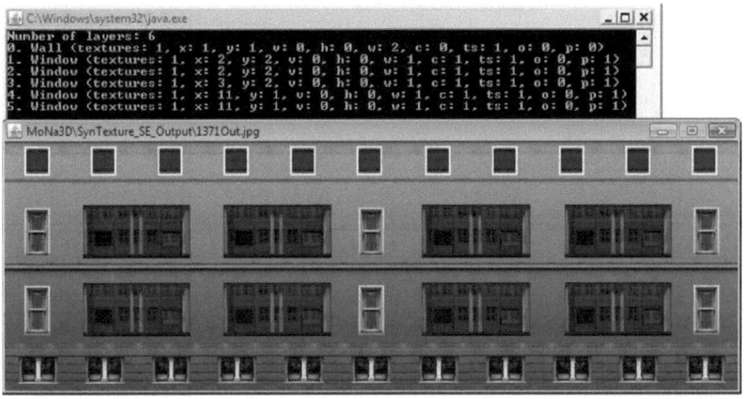

Figure 3: Running the JavaScript for generating the XML file as an input for Java program to generate the façade of HFT3 (Building 3 of Stuttgart University of Applied Science, Germany). The size of generated output image file is even smaller than the size of some of its own textures (747 * 256) 35.4KB and the quality is very high.

3.3 Dynamic 3D Navigation

Ivin et al. 2007 has designed the system as a background engine for the dynamic shortest path calculation of the 3D road network. Couples of values need to be inputted before running the program, start (origin) location and end (destination) location (see Figure 4).

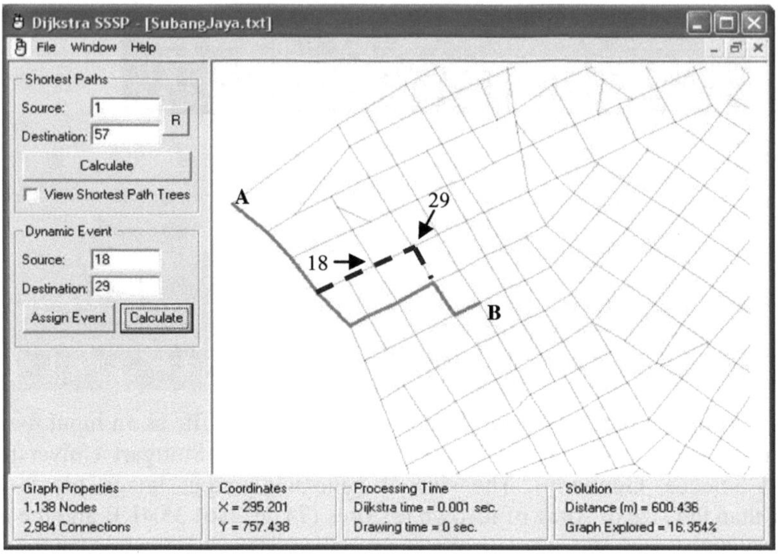

Figure 4: Shortest path from location A (vertex 1) to B (vertex 57) using Incremental SSSP Dijkstra algorithm (Ivin et al., 2007) after assigning a dynamic event at vertex 18 (source) and vertex 29 (destination). The dotted line was the original shortest path route from A to B before the dynamic event.

After calculating the results, it will be parsed to the newly developed SQL function (3D line buffering) to create a new focus map and in VRML file (*.wrl) for displaying shortest route within 3D city model.

3.4. The 3D Buffering

3D buffering zone could be defined as an operation to generate proximity information of a spatial object, for instance, phenomenon along linear features or polygon features in a three-dimensional space. The operation could be considered as one of the 3D analytical functions for 3D GIS. In GIS, such analytical tool is useful for analyzing spatial objects. In this section, the research deals with the development of the algorithms for 3D buffering tool as a framework for 3D analytical solution for creating focus map using 3D GIS approach. The module consists of the 3D geospatial primitive, i.e. line.

The buffering object will be studied in detailed (i.e. how to construct them geometrically) based on the geospatial primitive (line). To initiate the object model for buffering zone, the approach of the buffering object for the primitive will be mentioned. Later, all the mathematics that follows and the expected buffering result will be highlighted. On the other hand, the structures of each of the geospatial primitive need to be identified. This is due to the construction of the geospatial primitives affects the development of a buffering object model. Although the point and polygon buffering is out of the scope of this research, this paper will define the structural model on the buffering results of each of the geo-spatial primitives. Figure 5 shows the structural model of the 3D geo-spatial objects.

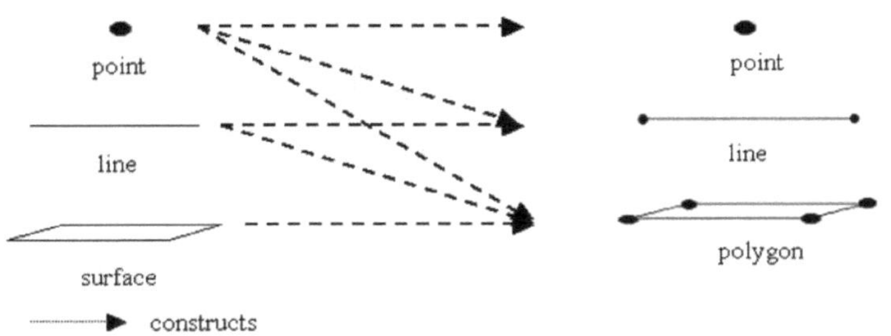

Figure 5: Geospatial object.

From the Figure 5, a polygon is formed by points, lines, and its surface, whereas a line consists of points, and the line itself. However, a point remains the same structure. After identifying the formation of each of the geospatial primitives, the buffering object modeling are easier to be performed since the scope of the modeling are narrowed into three buffer model. There are point (for point, line, and polygon), line (for line and polygon), and surface (for polygon) buffer. Yet, the final buffering model for the geospatial primitives are not that superficial as mentioned before. Therefore, the detailed studies about the solid buffer model for line will be discussed later in this chapter.

3.4.1. 3D Line Buffering

Two end nodes together with zero or more internal nodes define a line. Another phrase like polyline, arc or edges is also being used in GIS. Rolf (2000) had mentioned line data are used to represent one-dimensional objects such roads, railroads, canal, rivers and power lines. Refer to the Figure 4; line is a combination of both nodes and edge. The straight parts of a line between two successive vertices (internal nodes) or end nodes are called line segments. Therefore, in order to line buffer in 3D space, the joined components of a single line need to be identified (i.e. the nodes and edge(s)), see Figure 6.

Figure 6: Line's properties.

As mentioned earlier, point buffering output is a sphere, whereas the line buffering output is a cylinder. Since nodes appear in the line segment, adding the point buffering zone into the line buffer model becomes a necessity. Figure 7 shows the method to create a line buffering output. Details on the mathematical computational for line buffering can be referred to Chen et. al. (2007b).

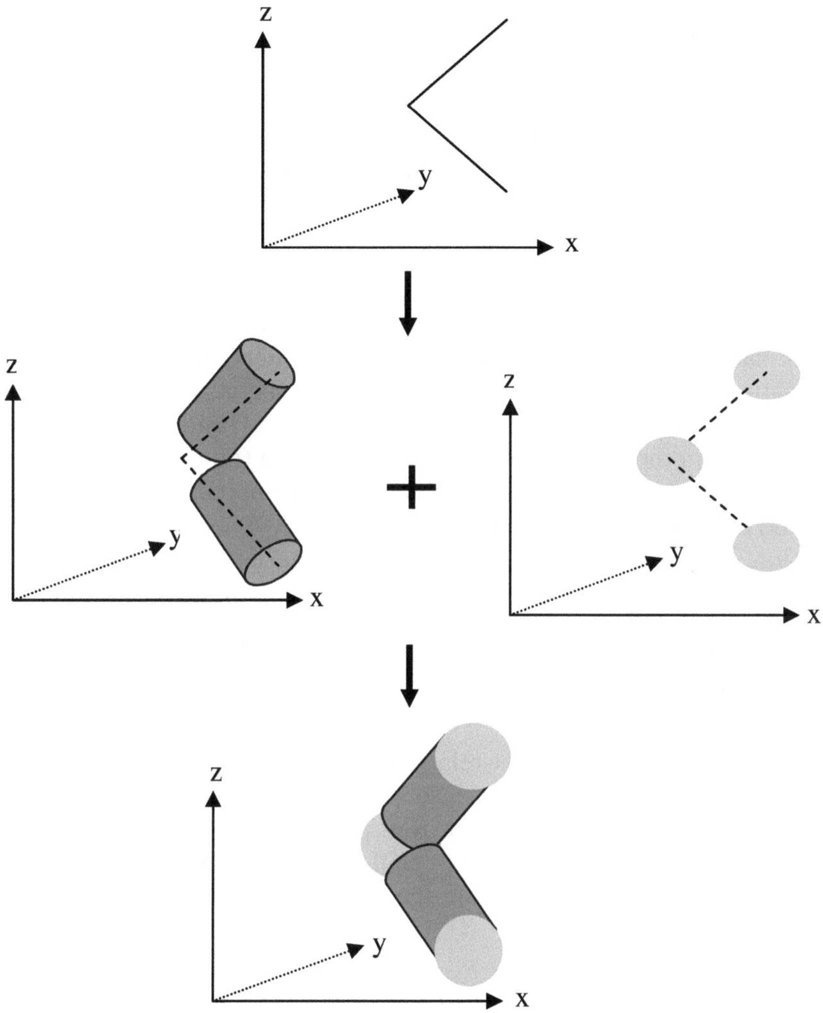

Figure 7: Method to create line buffer.

4. The Implementation

Existing DBMS provides a SQL schema and functions that facilitate the storage, retrieval, update, and query of collections of spatial features. Most of the existing spatial databases support the object-oriented model for representing geometries. The benefits of this model are that it supports for many geometry types, including arcs, circles, and different kinds of compound objects. Therefore, geometries could be modeled in a single row and single column. The model also able to create and maintain indexes, and later on, perform spatial queries efficiently. In the next section, some commercial spatial database will be discussed, in term of their characteristics, capabilities and limitations in handling multi-dimensional datasets.

4.1. Oracle Spatial

Oracle Spatial is designed to make spatial data management easier and more natural to users of location-enabled applications and geographic information system (GIS) applications. Once spatial data is stored in an Oracle database, it can be manipulated, retrieved, and related to all other data stored in the database. Types of spatial data (other than GIS data) that can be stored using Spatial include data from computer-aided design (CAD) and computer-aided manufacturing (CAM) systems. The Spatial also stores, retrieves, updates, or queries some collection of features that have both non-spatial and spatial attributes. Examples of non-spatial attributes are name, soil_type, landuse_classification, and part_number. The spatial attribute is a coordinate geometry, or vector-based representation of the shape of the feature.

Spatial supports the object-relational model for representing geometries. The object-relational model uses a table with a single column of SDO_GEOMETRY and a single row per geometry instance. The object-relational model corresponds to a "SQL with Geometry Types" implementation of spatial feature tables in the Open GIS ODBC/SQL specification for geospatial features.

4.2. Modeling 3D object using multipolygon

In the Oracle Spatial object-relational model, a 3D solid object from 3D primitive is possible, e.g. polyhedron. However, the study selected multipolygon to construct a 3D spatial object within geo-DBMS environment. This is due to two reasons: (i) the implementation of polyhedron from Oracle Spatial 11g could yield additional data structure for spatial object. The data structure of polyhedron define 3D object as a macro in the first place, e.g. POLYHEDRON { }. Then, each of the faces will be defined within the macro of polyhedron, e.g. POLYHEDRON {(Face1), (Face2), (Face3)…}. Finally, the vertices that construct each of the faces will be inserted within the data structure of faces. Comparing the multipolygon, the data structure that involves macro and micro will not be used. The implementation of multipolygon could yield a simple data structure that defines a 3D spatial object. (ii) Taking the advantage of 3D visualization, the study integrates with 3D display tool that support up-to multipolygon for 3D spatial objects. It could be done by implementing the multipolygon that bound a solid. The geometric description of a spatial object is stored in a single row and in a single column of object type SDO_GEOMETRY in a user-defined table. Any tables that have a column of type SDO_GEOMETRY must have another column, or set of columns, that defines a unique primary key for that table. Tables of this sort are sometimes referred to as geometry tables.

Oracle Spatial defines the object type SDO_GEOMETRY as:

```
CREATE TYPE sdo_geometry AS OBJECT (
SDO_GTYPE NUMBER,
SDO_SRID NUMBER,
SDO_POINT SDO_POINT_TYPE,
SDO_ELEM_INFO MDSYS.SDO_ELEM_INFO_ARRAY,
SDO_ORDINATES MDSYS.SDO_ORDINATE_ARRAY);
```

An example implementing 3D Multipolygon (where the geometry can have multiple, disjoint polygons in 3D) is given as below:

```
CREATE TABLE Solid3D (
   ID number(11) not null,
   shape mdsys.sdo_geometry not null);

INSERT INTO Solid3D (ID, shape) VALUES (
   1 SDO_GEOMETRY(3007,    -- 3007 indicates a 3D multipolygon
   NULL,                   -- SRID is null
   NULL,                   -- SDO_POINT is null
   SDO_ELEM_INFO_ARRAY (         -- the offset of the polygon
      1, 1003, 1,
      16, 1003, 1,
      31, 1003, 1,
      46, 1003, 1,
      61, 1003, 1,
      76, 1003, 1),
   SDO_ORDINATE_ARRAY (
      4,4,0, 4,0,0, 0,0,0, 0,4,0, 4,4,0, -- 1st lower polygon
      4,0,0, 4,4,0, 4,4,4, 4,0,4, 4,0,0, -- 2nd side polygon
      4,4,0, 0,4,0, 0,4,4, 4,4,4, 4,4,0, -- 3rd side polygon
      0,4,0, 0,0,0, 0,0,4, 0,4,4, 0,4,0, -- 4th side polygon
      0,0,0, 4,0,0, 4,0,4, 0,0,4, 0,0,0, -- 5th side polygon
      0,0,4, 4,0,4, 4,4,4, 0,4,4, 0,0,4  -- 6th upper polygon
)));
```

The advantage of implementing the multipolygon in DBMS is that the integration between CAD and GIS is possible for 3D visualization, i.e. Oracle (or called Spatial) spatial schema is supported by Bentley MicroStation (2008) and Autodesk Map 3D (2009). This is due to the geometry column provided by Spatial is directly access to the 3D coordinates of the object, which allow the display tools retrieve spatial information from the geometry column. The overall process of modeling visibility through visual landmarks can be described as follows:

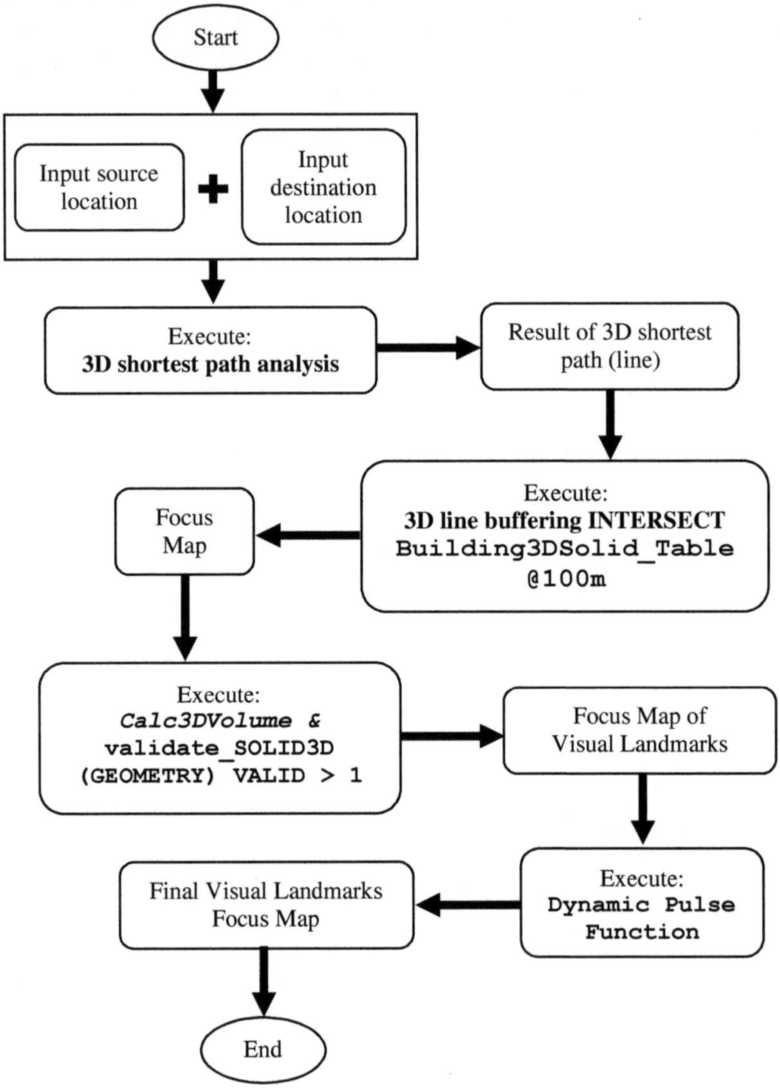

Figure 8: Process of generating visual landmarks using Geo-DBMS.

5. Experiment and Discussions

In order to provide a practical solution for the proposed new approach as discussed at previous sections, this section will discuss the current geo-DBMS implementation. The approach started with validation of input data into spatial DBMS. Although the 3D spatial object created for the study implements *multipolygon* that construct a solid object, some characteristics of 3D object need to be defined in order to create a valid object within geo-DBMS environment. The characteristics of valid object could be found in Aguilera & Ayala (1997); Aguilera (1998). The implementation of such validation within geo-DBMS environment could be found in Chen *et al.* (2007a); Chen *et al.* (2007a).

5.1. Object Validation

It is important that the spatial data is checked when it is inserted, or updated correctly into the DBMS. This check on the geometry of the spatial objects is called validation. Valid objects are necessary to make sure the objects can be manipulated in a correct way. The validation is required for dataset insertion, e.g. for point, line, polygon, polyhedron. It can be done using the SQL statement within the SQL terminal. For the validation functions for point, line, face, and solid object, only input data could be done within the SQL terminal. The validate process each type of data types is based on the rules for object construction. For 3D solid object similar to polyhedron, the validation rules are similar to the rules given by Arens (2003). A polyhedron is valid if;

1. The data structure is stored correctly;
2. Each polygon is a flat faces;
3. It bounds a single volume;
4. Its faces are simplicity, i.e. edges are not intersecting each others;
5. Vertices are stored as a face is correctly orient-able.

Each validation function can be implemented only one rule mentioned above. However, for the sake of simplification, the study combines all the rules into single validation function. Such validation is created as user-defined function and mapped into Oracle Spatial. The following SQL statement denotes the dataset within geo-DBMS and validation function within DBMS environment is given as follows.

```
SELECT validate_SOLID3D(GEOMETRY) VALID from test where
PID=1;
```

The result:

```
PID   VALIDATION_SOLID3D
---------------------
 1     TRUE
```

5.2. Dataset insertion within geo-DBMS

The implementation of geo-DBMS approach for managing disaster using 3D GIS approach starts with the database creation. In the study, each geometry table of point, line, face, and solid object was created in order to store the spatial object (see Figure 9). The GEOMETRY column stores the important coordinate structure of 3D spatial objects. The geometry information could be found in Figure 10. Some of the topological data structures are stored explicitly within each of the geometry tables, e.g. point-line, line-face, and face-solid relationships. The intention of storing both geometry and topological structures is to provide visualization (from geometry column) and spatial query (from topological structure) purposes.

```
SQL> DESC SOLID_3D;
 Name                                       Null?    Type
 ----------------------------------------- -------- ----------------------------
 SOLID_ID                                   NOT NULL VARCHAR2(50)
 SOLID_POLYGON                                       VARCHAR2(50)
 GEOMETRY                                            MDSYS.SDO_GEOMETRY
```

Figure 9: Geometry table for solid

```
SQL> SELECT * FROM SOLID_3D;

SOLID_ID                                           SOLID_POLYGON
-------------------------------------------------- --------------------------------------------------
GEOMETRY(SDO_GTYPE, SDO_SRID, SDO_POINT(X, Y, Z), SDO_ELEM_INFO, SDO_ORDINATES)
--------------------------------------------------------------------------------------------------------------
S01                                                P01,P02,P03,P04,P05,P06
SDO_GEOMETRY(3007, NULL, NULL, SDO_ELEM_INFO_ARRAY(1, 1003, 1, 16, 1003, 1, 31, 1003, 1, 46, 1003, 1, 61, 1003, 1, 76, 1
003, 1), SDO_ORDINATE_ARRAY(17.027, 39.9499, 4.2, 21.0583, 38.1107, 4.2, 22.7891, 41.9939, 4.2, 18.7372, 43.7766, 4.2, 1
```

Figure 10: Geometry table for solid

The 3D buffering function (line type) has been created as a new focus map based on the resulting 3D shortest path route analysis to determine which building will be used for generating measurable visual landmarks using dynamic pulse function. The implementation could be found at Section 5.3 (see Figure 12).

```
SELECT FunctionName (Face1, Face2, Face3, Facen, Buffer
Radius) FROM SolidTableName;
```

5.3. 3D Visualization of Visual Landmarks

3D visualization is an important phase that helps in presenting the output data (visual landmarks). Geospatial data is quite different from other kinds of data, especially for multi-dimensional geometry data structures and topological relationship. Current computer graphic technologies have considerable potential to extend the power of geo-spatial data visualization methods which traditional GIS can not provide. In this paper, VRML is used to visualize the output from executing newly developed Geo-DBMS SQL function in Oracle.

In the study, 3D city model of Stuttgart (see Figure 11) was used. It comprises of basic solid building structures in level of detail (LOD) 1 coordinated on draped aerial photograph with Digital Terrain Model (DTM).

Figure 11: 3D Model Stuttgart, © Stadtmessungsamt Stuttgart

The following SQL statement will be executed as background process. It denotes the 3D line buffering function for determining buildings that are intersected with the 100m focus map zone (cylinder type object obtained from resulting 3D shortest path route) from table Building3DSolid. The outputs are the intersected faces that form each of every building in the city model. Based on the Building_ID attribute, next step is to generate photo realistic façade of the selected buildings using dynamic pulse function.

```
SELECT LineBuffer3D_INTERSECT_Solid3D(F091, F092,
F100, F101, F113, F114, F430, F431, 100.00) FROM
Building3DSolid_Table;
```

The result:

```
 SID   BUILDING_ID
---------------------
 1     101
 2     102
 3     237
 4     236
```

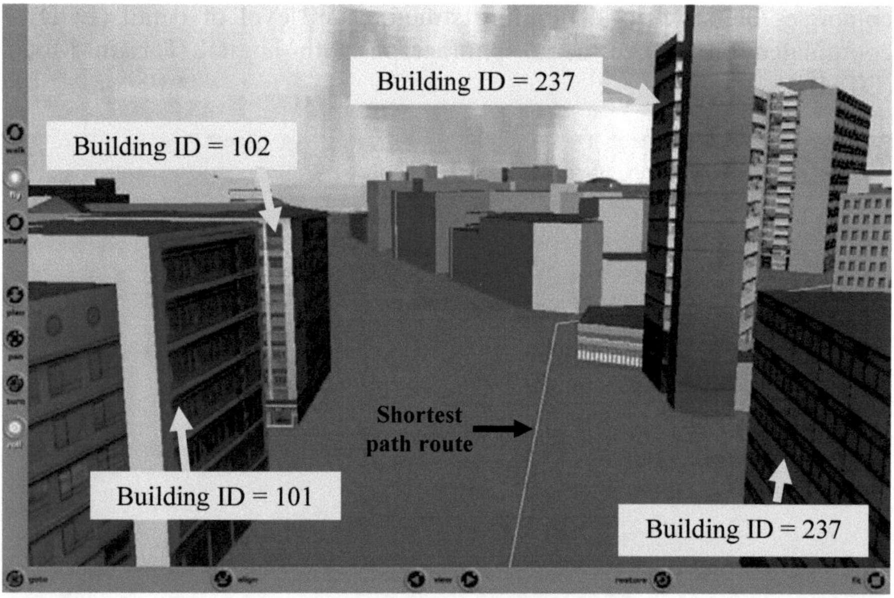

Figure 12: Final Visual Landmarks Focus Map of Stuttgart.

Figure 12 shows the result of visual landmarks visibility using Geo-DBMS approach. By implementing measureable visual landmark criteria (largest volume, tallest building and complex structure) only certain buildings in the focus map (obtained from intersection function between 3D line buffering and Building3DSolid_Table) are being mapped its detailed façade. The rest are only shown as LOD1.

6. Concluding Remarks

We have implemented an approach for generating measurable visual landmarks in a typical 3D city model based on couple of combined approaches: (1) dynamic pulse function (for generating detailed photo realistic building façade), (2) two 3D analytical operations; (i) 3D buffering (for generating focus map area from 3D shortest path analysis result) and (ii) 3D shortest path analysis for finding optimal route in 3D navigation in a city model. The proposed measurable criteria of visual landmarks in a 3D city model are based on top ranking of (1) largest volume of buildings, (2) most complex of building structure and (3) tallest building. The results have shown that the implementation of 3D buffering (3D line buffer intersect with 3D building solid) using 3D data type allowing 3D GIS analysis practical for such 3D application. Our concept was tested within Oracle Spatial 11g computing environment and has provided a promising outcome with respect to the developed algorithms.

In terms of 3D visualization of landmarks, by using the proposed dynamic pulse function for generating calculated layers of building's façade objects (windows, doors, corridors, etc.) the final output of the building model will result in high quality with low image file size, thus will provide better loading time at the end user side. If this is to compare with the used of LoD for visualization, the use of image texturing especially in LoD 2 and 3 will took a lot of effort and will increased the loading time, due to bigger image file size as far as the image quality is concerned. Therefore the use of dynamic pulse function will be most likely benefits to mobile navigation applications.

Future work will be specifically focused on the interaction between 3D city map information and Geo-Services for the mobile service of object recognition or awareness. Object awareness provides a concept for semantic indexing into huge information spaces where standard approaches suffer from the high complexity in the search processing otherwise, thus providing a means to relate the mobile agent to a semantic aspect of the environment. Visual object recognition using innovative and robust pattern recognition methodologies is an emerging technology to be applied in mobile computing services. Due to the many degrees of freedom in visual object recognition, it is highly mandatory to constrain object search by means of context based attention.

In contrast to signal based approaches as proposed in most location based services, object awareness points towards semantic indexing which will enable much more flexible interpretation of the information from a local environment. Object awareness relies on a structural matching process of comparing the situated information extracted from the sensor streams with prototypical patterns that were developed from experience. In the case of computer vision, object recognition methods enable to identify characteristic patterns of visual information in the field of view, such as, infrastructure objects (buildings, tourist sights, traffic signs, information boards, etc).

References

Aguilera, A, and Ayala, D. 1997. Orthogonal polyhedra as geometric bounds in constructive solid geometry. In C. Hoffman, and W. Bronsvort, (eds.), *Fourth ACM Siggraph Symposium on Solid Modeling and Applications*, Vol.4: 56-67.

Aguilera, A 1998, Orthogonal polyhedra: study and application. *Ph.D. Thesis*, LSI-Universitat Politècnica de Catalunya.

Arens, C. A. 2003. Modeling 3D spatial objects in a geo-DBMS using a 3D primitives. Msc thesis, TU Delft, The Netherlands. 76 p.

Chen TK, Abdul-Rahman A, Zlatanova S (2007a). New 3D data type and topological operations for geo-DBMS. In: V. Coors, M. Rumor, E. Fendel, and S. Zlatanova (eds.): Urban and Regional Data Management, UDMS Annual 2007, Taylor Francis Group, London, pp. 211–222.

Chen TK, Abdul-Rahman A, Zlatanova S (2007b). 3D spatial operations in geo-DBMS environment for 3D GIS. In: O. Gervasi and M. Gavrilova (eds.): ICCSA 2007, LNCS 4705, Part I. Springer-Verlag, Berlin. pp. 151–163.

C. Brenner and B. Elias, 2003. Extracting Landmarks for Car Navigation Systems Using Existing GIS Databases and Laser Scanning. Proceedings of Photogrammetric Image Analysis, International Archives of the Photogrammetry, Remote Sensing and Spatial Information Sciences, Vol. XXXIV, Part 3/W8, 2003.

Coors, V. & A. Zipf, 2007. MONA 3D - Navigation Using 3D City Models. Available at: http://www.geographie.uni-bonn.de/karto/MoNa3D.Telecarto-2007-vc.az.full.pdf

Deakin, A., 1996: Landmarks as Navigational Aids on Street Maps. Cartography and Geo-graphic Information Systems, 23 (1): 21-36.

Denis, M.; Pazzaglia, F.; Cornoldi, C.; Bertolo, L., 1999: Spatial Discourse and Navigation: An Analysis of Route Directions in the City of Venice. Applied Cognitive Psychology, 13: 145-174.

GTA Geoinformatik GmbH - http://www.gta-geo.com/eng/index_e.html

Golledge, R. D., 1999. Wayfinding Behavior. John Hopkins Press, chapter Human Wayfinding and Cognitive Maps, pp. 5– 45.

Ivin Amri Musliman, Alias Abdul Rahman and Volker Coors, 2007. Modeling Dynamic Weight for 3D Navigation Routing. Proceedings of Map Asia 2007, 14-16 August, Kuala Lumpur

Langdon, D., Soto, A., and Mery, D. 2006. Automatic Selection and Detection of Visual Landmarks Using Multiple Segmentations. In: L.-W. Chang, W.-N. Lie, and R. Chiang (Eds.): PSIVT 2006, LNCS 4319, pp. 601–610, 2006.

Lynch, K., 1960: The Image of the City. MIT Press, Cambridge, 194 pp.

Michon, P.-E.; Denis, M., 2001: When and Why are Visual Landmarks Used in Giving Directions? In: Montello, D. (Ed.), Spatial Information Theory. Lecture Notes in Computer Science, 2205. Springer, Berlin, pp. 292-305.

M. Schulze-Horsel, 2007. Generation, Characteristics and Applications Conference – 3D Architecture 2007.

Neis, Pascal and Zipf, Alexander, 2007. Realizing Focus Maps with Landmarks using OpenLS Services. Available at: www.geographie.uni-bonn.de/karto/LBSTelecarto2007.FocusMapsWithOpenLS.last.pdf

Olivier Stasse, Andrew J. Davison, Ramzi Sellaouti and Kazuhito Yokoi, 2006. Real-time 3D SLAM for Humanoid Robot considering Pattern Generator Information. Available at http://ieeexplore.ieee.org/stamp/stamp.jsp?arnumber=04058956

Raubal, M., & Winter, S., 2002. Enriching Wayfinding Instructions with Local Landmarks. In Geographic Information Science. (Egenhofer, M.J., & Mark,

D.M., eds.), Lecture Notes in Computer Science, Vol. 2478, Berlin, Springer, pp: 243-259.

Rolf, A. de By (2000). "Principles of Geographic Information Systems." ITC, The Netherlands, 230 p.

S. Se, D. Lowe, J. Little, 2001. Local and Global Localization for Mobile Robots using Visual Landmarks. Proceedings of IEEE/RSJ International Conference on Intelligent Robots and Systems, IROS 2001, pages 414-420, Maui, Hawaii, October 2001.

Sorrows, M. and Hirtle, S., 1999. The nature of landmarks for real and electronic spaces. In: C. Freksa and D. Mark (eds), Spatial Information Theory: Cognitive and Computational Foundations of Geographic Information Science, Springer Verlag, pp. 37–50.

Swienty, O. & Reichenbacher, T. (2006): Relevanz und Kognition in der mobilen Kartographie. Aktuelle Entwicklungen in Geoinformation und Visualisierung. GEOVIS 2006, Germany, GFZ Potsdam, April 5–6.

Tversky, B. and Lee, P., 1999. Pictorial and verbal tools for conveying routes. In: C. Freksa and D. Mark (eds), Spatial Information Theory: Cognitive and Computational Foundations of Geographic Information Science, Springer Verlag, pp. 51–64.

Vande Velde, L., 2004. Navigation in 3D, Enhanced map display for car navigation and LBS, Proc. of 11th World Congress on ITS, ITS 2004, Nagoya, Japan, 2004.

Werner, S.; Krieg-Brückner, B.; Mallot, H.; Schweizer, K.; Freksa, C., 1997: Spatial Cognition: The Role of Landmark, Route, and Survey Knowledge in Human and Robot Navigation. In: Jarke, M.; Pasedach, K.; Pohl, K. (Eds.), Informatik '97. Springer, pp. 41-50.

Zhao, Y., 1997. Vehicle Location and Navigation Systems. Artech House, Inc. Boston, London.

Zipf, A. (2002): User-Adaptive Maps for Location-Based Services (LBS) for Tourism. In: K. Woeber, A. Frew, M. Hitz (eds.), Proc. of the 9th Int. Conf. for Information and Communication Technologies in Tourism, ENTER 2002. Springer, Berlin.

Zipf, A. and Richter, K.F., 2002. Using Focus Maps to Ease Map Reading. Developing Smart Applications for Mobile Devices. In: Künstliche Intelligenz (KI) (Artificial Intelligence). Special Issue: Spatial Cognition. 04/2002. 35-37.

A 3D inclusion test on large dataset

Kristien Ooms, Philippe De Maeyer and Tijs Neutens

Department of Geography, Krijgslaan 281 S8, 9000 Ghent, Belgium

Abstract. This paper proposes a 3D inclusion test which enables processing very large datasets in a limited amount of time. Each point in the dataset is provided with a label according to its location relative to a polyhedron. These 3D objects can be modeled in a number of ways, related to its purpose or application in which it is used. The two representation types considered in this paper are the Boundary Representation (B-Rep) and the Constructive Solid Geometry (CSG). In order to be able to process a large amount of points in a relative limited amount of time, a preprocessing phase is essential prior to performing a sequence of operations. In this initial step, the space is subdivided into a 3D array of voxels. The voxelization procedure deviates slightly between the two representation types, but the general principle remains the same. This subdivision allows fast retrieval of the status (inside/outside) of a very large amount of points in space.

1 Introduction

An inclusion test is an indispensable base operation in a GIS environment which users often employ to verify whether or not a point is located within another object. The test is also an essential starting point for more complex operations. As a consequence, the efficiency of this method influences the overall efficiency of frequently used operations. This is especially true when analyzing large datasets. Using an unsuitable method to process these data would result in unacceptable processing times. Special methods are required to handle these large datasets, avoiding unnecessary repetition of computational intensive algorithms.

T. Neutens, P. De Maeyer (eds.), *Developments in 3D Geo-Information Sciences*,
Lecture Notes in Geoinformation and Cartography, DOI 10.1007/978-3-642-04791-6_10,

So far, little attention has been paid to inclusion tests on 3D volumes. For points and polygons in a 3D space, the extension of the algorithms is rather straightforward by projecting the polygons and points onto a surface and testing them in 2D (Taylor 1994). Taking into account the 3D characteristics of objects, more fundamental adaptations of existing 2D methods are needed or new algorithms have to be developed. The method proposed by Feito and Torres (1997), for example, is such a 3D extension of the 2D triangular method mentioned earlier. Another example of a specialized 3D inclusion test is proposed by Walker and Snoeyink (1999) using the CSG-tree. A major drawback of these tests is that they are not suitable for handling large datasets.

Hence, the purpose of this paper is the development of a 3D inclusion test in the third dimension which can process very large datasets within acceptable time limits. The method is capable of working with two different types of object models most often used in a geographical context: the Boundary Representation and the Constructive Solid Geometry (see section 2.2). As a consequence, the results from the inclusion test can be compared in a uniform way, although the objects may originally be modeled in a different type. This also facilitates further analyses based on the inclusion test.

The paper is organized as follows. In the next section (section 2.1), a literature overview of the inclusion test is presented. The need for a preprocessing phase is also explained in this section. In section 2.2, the two main 3D object representations are described: the Boundary Representation and the Constructive Solid Geometry. Next, in section 2.3, a promising 2D inclusion test, proposed by Gombosi and Zalik (2005) and Zalik and Kolingerova (2001), is described. This test will form the basis for the 3D inclusion test proposed in this paper. Then, the different steps of the 3D inclusion test are presented in detail (see section 3). In these steps, the adjustments related to the different object representations – B-Rep and CSG – are emphasized. Finally, a comparison is made between the different methods.

2 Background

2.1 Inclusion tests

The main principle of the inclusion test is rather simple: one wants to test whether or not a point $p(x_p,y_p,z_p)$ is located within another object. The type of this object depends on the dimensions of the environment (1D, 2D, 3D). The inclusion tests in a 2D space can be based on a wide range of techniques which all have their advantages and disadvantages towards certain special cases in polygon design. Basically, these tests can be subdivided into two main categories depending on whether or not a preprocessing phase is used. The best known methods, which do not use this preprocessing phase, include ray crossing algorithms (Gombosi and Zalik 2005; Goodchild 1990), angular methods (Haines 1994; Preparata and Shamos 1985), the sign of offset method (Huang and Shih 1997; Taylor 1994; Zalik and Kolingerova 2001) and the method of the winding numbers (Hormann and Agathos 2001). The time complexity for testing a single point with this type of methods is always $O(n)$, with n the number of vertices of the polygon (Gombosi and Zalik 2005; Huang and Shih 1997; Li et al. 2007; Zalik and Kolingerova 2001).

However, when a large number of points need to be tested, the fact that each point has to pass separately through the method increasingly becomes computationally prohibitive. A method with a preprocessing phase may be more justified in such a case. This preprocessing phase is the most computationally intensive part of the method, but is executed only once. The results of this initial phase are essential to allow a fast examination of large datasets. During the preprocessing phase, the objects are decomposed into certain elements which are more easily processed during the actual inclusion method. These elements are for example grid cells (Gombosi and Zalik 2005; Zalik and Kolingerova 2001), the polygon's edges (Wang et al. 2005), triangles (Feito et al. 1995; Haines 1994) or a set of convex polygons (Li et al. 2007).

Regarding algorithms of the second category, the complexity of the preprocessing phase has to be taken into account besides the complexity of the actual inclusion test. In most cases, this initial phase can be finished in a linear time, $O(n)$, with n also the number of vertices. The subsequent inclusion test, on the other hand, is realized in a shorter time than $O(n)$, such as $O(\log n)$ or even in a constant time $O(1)$. (Gombosi and Zalik 2005; Huang and Shih 1997; Li et al. 2007; Zalik and Kolingerova 2001)

In Fig. 1 a general comparison between the CPU times needed to test a number of points for inclusion is presented. The number of points to be tested is presented on de X-axis with a logarithmic scale. The dashed line in the graph depicts the CPU times needed for an algorithm with a preprocessing phase. In this case, a time complexity of *O(n)* is assumed for the preprocessing phase and *O*(1) for the testing phase. The number of vertices is considered a constant in this graph. Without the preprocessing phase, the CPU time needed to perform the test rises in an exponential way. The steepness of the function line linked to the method with the preprocessing phase on the other hand is much less, resulting in more acceptable processing times in the case of many test points. The figure thus clearly illustrates the advantage of the preprocessing phase when large datasets need to be tested.

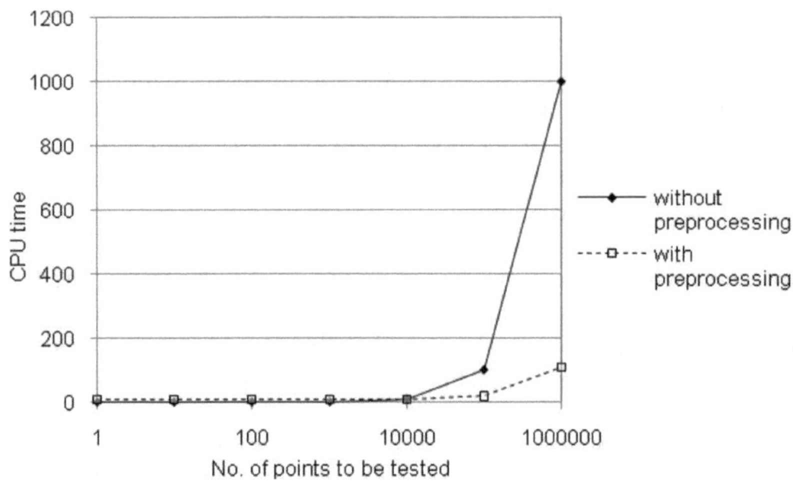

Fig. 1. : Relation between the processing time and the size of the dataset, with and without processing

The algorithm proposed in this paper is a 3D extension of the Cell Based Containment Algorithm (CBCA) by Zalik and Kolingerova (2001). Therefore, the most commonly used 3D object models in a GIS and CAD environment are discussed in the next section (section 2.2): the Boundary Representation and the Constructive Solid Geometry. In section 2.3, a short description of the 2D CBCA method is given.

2.2 3D object modeling: B-Rep and CSG

When the inclusion method is extended to the third dimension, a new object needs to be considered in the tests: the volume. Such a volume can be modeled in different ways according to the application in which it is used. Consequently, these different modeling techniques influence the exact implementations of the inclusion methods for the third dimension. In this paper two such models, often used to describe 3D objects, are considered: the Boundary Representations (B-Rep) and the Constructive Solid Geometry (CSG).

Other types of 3D models exist to describe the objects, such as the tetrahedral model. The B-Rep and CSG are specially selected as they are the most commonly used 3D object representations in the field of the Geographic Information Systems (GIS) and Computer Aided Design (CAD). The point in polygon (or polyhedron in 3D) is closely linked with these two fields.

When working with a B-Rep, the objects (in *n*D) are modeled by its boundaries (in *n-1*D). This means that a polyhedron (3D) is bounded by polygons (2D), the polygons are bounded by lines (1D) and the lines are bounded by vertices (0D). This hierarchical organization can be implemented by various data structures and yields a good approximation of real world objects (Stoter and Zlatanova 2003). Another important element when considering inclusion tests is the fact that these Boundary Representations create 'empty' objects. This results in three types of topological locations: inside the object, outside the object and at the border of the object.

The Constructive Solid Geometry, on the other hand, works with solids: objects which are completely 'filled up'. A number of primitive objects – cubes, cylinders and spheres – are combined to construct more complex objects, using a set of Boolean operations such as *union* and *intersect*. This type of model originates from the field of CAD, but has its limitations when representing real world objects as the relations between the primitives become rather complex. (Stoter and Zlatanova 2003)

2.3 The 2D CBCA

The 2D CBCA, proposed by Gombosi and Zalik (2005) and Zalik and Kolingerova (2001), will be described in this section. This method will form the basis for the 3D CBCA proposed in this paper. This 2D inclusion test is selected for a number of reasons. Firstly, it has a preprocessing phase which is – as explained in section 2.1 – necessary to be able to process large amounts of points. Secondly, the partitioning method used in the 2D CBCA, a grid, can be extended to the third dimension in a rather straightforward way: an array of voxels. Lastly, extending this method to the third dimension does not necessary require specific graphical hardware, which makes it possible to execute the method on a 'conventional' desktop PC.

This 2D algorithm can be subdivided into a number of subsequent steps. The base idea behind the algorithm is rather simple. In the preprocessing phase, a raster is constructed in which each cell contains information about its location: inside, outside or at the border of the polygon. In the actual (inclusion) test, the status of the cell – in which each point is situated – is derived (Gombosi and Zalik 2005).

Hence, a uniform grid is defined in the initial phase whereby the choice of the grid resolution is essential. This choice is determinative for the initialization time, amount of memory needed, processing time and accuracy of the method. Next, cells containing a part of the polygon border are determined and a flood fill algorithm is used to assign a value (outside/inside) to the remaining pixels. (Gombosi and Zalik 2005; Zalik and Kolingerova 2001)

In the second step, the algorithm verifies in which cell each point is located. The status of the corresponding cell thus holds the information about the location of the point. Only when a point is located in a cell with a border-status, further calculations are required. These calculations can be executed in detail or rather approximate. The approximate methods define cells which are for more than 50% located within the polygon as 'inside'; all the other cells are labeled 'outside'.

3 A 3D extension of the CBCA

The proposed method is a 3D extension of the Cell Based Containment Algorithm, described in the former sections. This 2D algorithm is considered to be the most suitable because it is capable to handle very large datasets, which is also the aim of this paper.

In the preprocessing phase, 3D space is subdivided into voxels. The method used to achieve this subdivision is dependent on the definition (representation) of the objects in the 3D space: B-Rep or CSG.

This grid of voxels can be seen as a spatial index allowing a regular 3D tessellation of space in rows, columns and levels. Furthermore, each element in the index is provided with a status linked with the purpose of the final query (inside/outside). The preprocessing phase is necessary to construct the spatial index (3D grid) and especially to be able to assign a status to each of the elements.

Besides the regular tessellation in a 3D grid, other structures of spatial indices can be considered such as the hierarchical structure of the octree or the R-tree. These more complex structures are not necessary to speed up the calculations during the actual inclusion test since a comparison between the coordinates of the points and the voxels are made to locate their mutual position. Moreover, the R-tree does not provide a regular tessellation of the space, which would complicate the calculations instead of causing improvement.

In the next sections, the proposed method is described step by step and at certain points, the deviations in the algorithm, due to the different modeling types, are stressed. An overview of the complete method can be found in Fig. 3 and Fig. 4 at the end of section 3.4.

3.1 Initialization

The first decision which has to be addressed is the choice of the voxel size. This initial choice has far going consequences in the next steps, influencing the initialization time, memory, processing time to test a point and accuracy in a significant way. In general, the size of the voxels in x, y and z does not have to be the same, but mostly a uniform subdivision with constant sizes is preferred. (Ogayar et al. 2005)

Three methods to define the voxel sizes are proposed. The first method has a fixed voxel size which will be applied to all scenes. A second, more flexible method, allows the user to determine the voxel size. This way the user has control over the accuracy (s)he needs and the associated processing times. A third method calculates an optimal voxel size dependent on the extent of the scene. The bounding box of the scene (around all the present objects) is constructed. The lengths of this box in x, y en z determine the optimal size of the voxels. A key parameter in these calculations is the maximal number of voxels which can be used in the process.

3.2 Voxelization: general requirements

After the voxel sizes are determined, the status of each voxel needs to be obtained. This status informs the user if a certain voxel is located inside or outside the object. The simplest voxel space is a 3D array in which a single bit is assigned to each voxel. The bit is activated (status = 1) if the voxel is inside the object and is, deactivated (status = 0) if the voxel is outside the object.

Two voxelization methods are described in the next sections, each related to the type of objects (B-Rep or CSG) to be voxelized. The selected methods also need to comply with a number of assumptions. Firstly, the voxelization phase needs to be limited in time, enabling fast results retrieval for the whole method. Secondly, it has to be possible to run the method on a PC without any extra (graphical) hardware requirements. Lastly, the amount of memory which needs to be allocated during the voxelization process has to remain limited. The blocky presentation of the objects is of minor importance because the main goal of the study is not focused on the visualization aspect.

The methods described in the next two sections comply with these assumptions. These methods are derived from existing studies (Fang and Chen 2000; Thon et al. 2004) and adapted where needed in order to fulfill the specific requirements, linked with our purposes, in a more optimal way.

3.2.1 Voxelization B-Rep

The Boundary-Representation is characterized by the objects' boundaries. As a consequence, the interior of the object is 'empty'. However, for the analysis it is essential that this empty interior is transformed into a set of voxels with status 'inside'. This fact needs to be kept in mind during the selection of a suitable voxelization algorithm.

Not all voxels in the 3D array will be located completely outside or inside the object. Certain voxels will contain the boundary of the object, which complicates the voxelization procedure. If only an approximated inclusion test is sufficient, voxels which are for more than 50% within the object are labeled 'inside' with value 1. If a more detailed test is desired, two bits need to be assigned to each voxel: 1 bit for the inside/outside question and one bit for the boundary/non-boundary question.

The voxelization algorithm which is considered to be the most suitable for these purposes is described in Thon et al. (2004). Below, the general structure of this method is outlined.

The method itself can be subdivided in two steps: a preparation step and the actual voxelization step. The method allows a uniform voxelization of 3D polygonal objects or polyhedrons. It is based on an optimized 'ray casting' method throughout the bounding surfaces of the object.

The first step in the voxelization process entails a subdivision of the space. This first step is necessary to speed up the calculations of the second step, where the intersection points between the 'rays' and surfaces are processed. The subdivision starts with the creation of a bounding box around the object, from which the surface parallel to XY plane is subdivided by means of a quadtree. Each node in this quadtree corresponds to a box parallel to the Z-axis, whose depth is equal to the depth of the bounding box. Each leaf of the quadtree contains a list of surfaces which are intersected by the box. (Thon et al. 2004)

The 3D grid, which was determined in the previous step, is placed over the object. Next, a ray is cast parallel to the positive Z-axis. This ray has its origin at the center of each cell of the uniform raster parallel to the XY-plane. Consequently, if the constructed grid consists out of m x n x p cells, only m x n rays are cast. (Thon et al. 2004)

For each ray – with for example (x_0, y_0, z_0) as origin – the corresponding box, and the corresponding leave in the quadtree is thus determined. Next, each surface listed in the leave is checked for intersection with the ray. When an intersection is found, its z-coordinate is placed in a linked list. This list is ordered according to a rising z-value. Consequently, the 3D grid is processed in a 2D way, cell by cell in the plane parallel to the XY-plane. (Thon et al. 2004)

The status of every voxel can be derived from the z-values stored in the different linked lists. The same rule used in the 2D ray casting method can be applied in this situation: the number of intersections between the ray and the object (a bounding surface of the polyhedron in this case) is counted. If this number is odd, the tested voxel is inside the object, otherwise it is located outside the object. The z-value of each voxel – linked with the same ray – is compared with the z-values in the linked list. If the number of z-values which are larger than the z-value of the voxel is odd, the voxel is located within the object. (Thon et al. 2004)

The z-values in the linked list are also useful to determine the voxels which contain the bounding surfaces of the objects. If the difference between the z-value of the voxels' centre and the z-value of an intersection is smaller than half the voxel size, then the bounding surface is located within the voxel.

3.2.2 Voxelization CSG

The 3D objects in the Constructive Solid Geometry are composed out of solids, combined by a number of Boolean operators. The interiors of these objects are thus filled. Furthermore, the selected voxelization method needs to be able to handle these combinations of primitives, which also makes it less easy to derive the bounding surfaces.

Fang and Chen (2000) proposed a method which can handle these constraints, but this method does not comply with the assumptions described earlier. It requires special graphical hardware which implies that it cannot be employed on a conventional desktop PC. The method proposed in the next paragraphs is an adaptation of the voxelization process used for the Boundary-Representation in such a way that it can handle the specific structure of the Constructive Solid Geometry. These adaptations are based on the main principles of the voxelization method described by Fang and Chen (2000).

The algorithm for the CSG is – like with the B-Rep – divided into two steps: a preparation step and the actual voxelization step. The preparation step is very similar to the preparation step for the B-Rep described earlier: a quadtree is built, based on the bounding box of the object, in which each leave corresponds with a box parallel with the Z-axis. The leaves contain a list with primitives (the composing elements for the 3D object in a CSG) which have an intersection with the box.

The subsequent step, the actual voxelization procedure, can be structured in two ways. Both methods are based on the combination of the primitives according to the defined Boolean operations in the CSG model. In the first method, the complete primitives are combined with each other, following the structure of the CSG tree. An example of the method is depicted in Fig. 2. Here, three primitives – a sphere, a cube, and a pyramid – are combined with the Boolean operation union (+) and difference (-). The tree is processed in a post order traversal. This way, two complete primitives are voxelized and the corresponding voxels are subsequently combined using their Boolean operator. This newly created voxelized object replaces the parent node of the two voxelized objects.

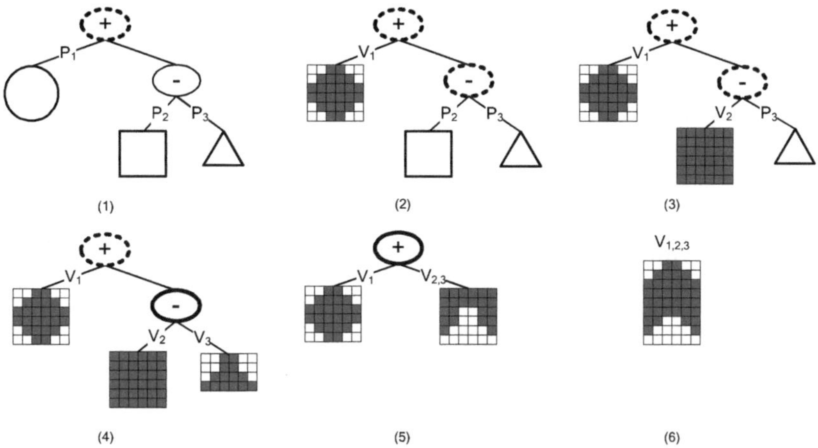

Fig. 2. : Processing of the CSG tree according to the first voxelization method

Each primitive is voxelized in an analogue manner as described earlier with a B-Rep model. The only difference here is that the lists in the leaves of the quadtree do not contain the bounding surfaces which have an intersection with the corresponding box; these lists now contain the primitives which are crossed by the ray. Thereafter, the intersections between the listed primitive(s) and the ray are determined and their z-values are stored in an ordered linked list.

Next, the status of all the voxels along the ray is derived by comparing the z-values of their centers with the ones stored in the linked list. If the number of z-values – with a higher value and linked to the same primitive – is odd, then the voxel is located within this primitive. In this case, the status of the voxel is 1 (inside) otherwise it is 0 (outside). Once two primitives, linked with the same parent node, are voxelized, they are combined according to the correct Boolean operator (Fig. 2).

The method described above is not very efficient in terms of computer memory, since two whole primitives need to be completely voxelized –

and thus stored in the memory – before the parent primitive can be computed. The main advantage of the method is that all the voxelized sub-complexes in the CSG tree are also computed and can thus be stored for other purposes. This may prove to be a critical asset when this point-in-object test is a starting point for more complex analysis.

A second variant of the method – which does not voxelize complete primitives – will be described in the next paragraph. Consequently, this method is to be preferred when the voxelized sub-complexes of the CSG are not needed in further analyses. This variant is closely related to the scan process, in which an image is created pixel by pixel. In this case, each ray – with all the primitives crossed by it – will be voxelized one after the other. The order in which the primitives in the ray are processed is the same as with the previous method, but the primitives will not be voxelized as a whole anymore. Only the part of the primitives in a certain ray is voxelized and subsequently combined according to the correct Boolean operators.

Hence, the scene is processed ray by ray; each ray creating a line of adjacent voxels which corresponds to the voxelization of the 3D object in that location. The entire 3D object is voxelized immediately after all the rays are processed.

3.3 Inclusion test

The actual inclusion test will check if a point is located in the interior or exterior of the object. The initial preprocessing step has divided space into a large number of elements (voxels) and the status (inside/outside) of each element is known. Hence, the aim of this second step is to determine in which voxel the points are located. The position of a voxel $v(r_v, c_v, l_v)$ in which a point $p(x_p, y_p, z_p)$ is located, can be derived by means of these formula:

$$r_v = \frac{x_p - x_{min}}{size_r}; \qquad c_v = \frac{y_p - y_{min}}{size_c}; \qquad l_v = \frac{z_p - z_{min}}{size_l}$$

The status of the voxel in which the point is located is thus the same as the status of the point. A problem arises when the voxel is located near the border of the object. Two methods of handling this situation can be distinguished: an approximate method and a detailed method. These methods are based on the work of Zalik and Kolingerova (2001).

3.3.1 Approximate inclusion test

The approximate inclusion test is the fasted method, but is also less accurate. In this case, the voxels can only have the status inside (value = 1) or outside (value = 0). One bit per voxel is thus sufficient to label these voxels, which reduces the memory needed to store the image.

At certain locations, the outer boundary of the object will pass through the voxels. A part of the voxel will thus be located inside the object and a part outside, but the voxel can only have the status inside or outside. This causes an error in the status of certain points located near the boundary of the object. This error rate is dependent on the accuracy of the voxelization and, consequently, on the resolution or size of the voxels.

The decision about the status of the boundary voxels is based on the amount of voxel space located within the object. If this amount is more than 50%, then the bit of the voxel is activated (value = 1) and otherwise it is deactivated (value = 0).

3.3.2 Detailed inclusion test

In this case, the user exactly wants to know if the point is located inside or outside the object. More information about the boundary voxels is required when this detailed test is desired. This means that two bits will be assigned to each voxel: one bit for the inside/outside status and one bit to point out the boundary voxels. This also gives extra information about the location of the boundary voxels: if a boundary voxels has the status 'inside', then this voxel is located for more than 50% in the object and vice versa.

This extra information requires more memory to store the voxelized image, but makes it possible to perform a detailed test. When the status of a voxel is 'boundary', an extra ray casting test is required. Information from the preprocessing step can be reused at this moment, since the ray casting method in the voxelization step is analogous to the method used in this step.

A ray is cast from the point which is tested, parallel with the positive Z-axis. This ray stops when it encounters a voxel with a status 'non-boundary'. To determine the exact status of the point, the number of intersections between the ray and the objects' boundaries are count. This information is contained in the quadtree constructed in the initialization step. If this number is odd, the status of the point is the opposite of the inside/outside status of the encountered voxel. This detailed test also allows determining if a point is located at the boundary of the 3D object.

3.4 A comparison between the different types

In the next paragraphs a comparison of the different types of inclusion tests is described. The four types considered are: a conventional test without preprocessing, the proposed inclusion test with a B-Rep object representation and the two variants of the inclusion test with the CSG object representation. An overview of the influencing factors and the predicted processing times are also given.

Firstly, the time complexity of the conventional inclusion tests – without a preprocessing phase – is related to the number of vertices in the object, thus *O(n)*. Consequently, with this type of algorithm every point to be tested has to pass through every step of the algorithm, resulting in rapidly rising processing times when large datasets are analyzed.

The processing times related to the algorithms proposed above are influenced by a number of factors, dependent on the object's representation type. The time needed for the last step of the algorithm, the actual inclusion test, is the same for the three algorithms and can be considered constant for each point. The main influencing factors on the processing times of the first phase are:

- the number of faces (or vertices) of the objects;
- the number of primitives (only for CSG);
- the number of voxels;
- the maximal depth of the quadtree.

An increase in most of these factors will also cause an increase in the processing time; only for the depth of the quadtree this is not completely true. The preprocessing phase is subdivided in a partition step and a voxelization step. Allowing a larger maximal depth for the quadtree means that the time needed to finish the partition step will increase. The processing time related with the actual voxelization step is not uniquely linked with the maximal depth of the quadtree. In this case, an optimal depth can be noticed. (Thon et al. 2004)

The weight of these factors on the processing time depents on the type of object representation (B-Rep or CSG). Although, the overall time complexity for the three methods (one for the B-Rep and two for the CSG) will be nearly the same. This would result in a similar graph as depicted in Fig.1. The number of points to be tested is presented on the X-axis, whereas all other factors are considered constant.

Although the voxelization method is not designed to work in real time, relative fast results can be achieved for the preprocessing phase. Thon et al. (2004) have discussed following processing times for the voxelization process of an object which consists out of 10444 faces and using a quadtree depth of 5, for a resolution of 64^3: 0,047s; 128^3: 0,172s; 256^3: 0,870s and 512^3: 6,766s. This resolution is related to the number of voxels used to voxelize the object. For example, with a resolution of 64^3 the object is voxelized using an array of 2621444 voxels: 64 voxels along each axis. Using more voxels results is smaller voxels and thus in a more accurate approximation of the object. This, however also causes the processing time to rise, both for creating and for querying the array

Computation times related to the voxelization process of objects modeled in a CSG, on the other hand, will be longer since each individual primitive of the CSG tree needs to be considered. This is also influenced by the option chosen for the voxelization procedure of the CSG-object. More individual objects need to be voxelized, but since these objects are primitives, their general shape is much simpler than the shape of a B-Rep: they have fewer faces and thus fewer vertices which reduces the processing times.

The actual inclusion tests can be finished in constant time, *O(1)*, for each point. Consequently, very large datasets can be tested in a very small amount of time as the calculation times of the preprocessing phase are independent on the number of points to be tested.

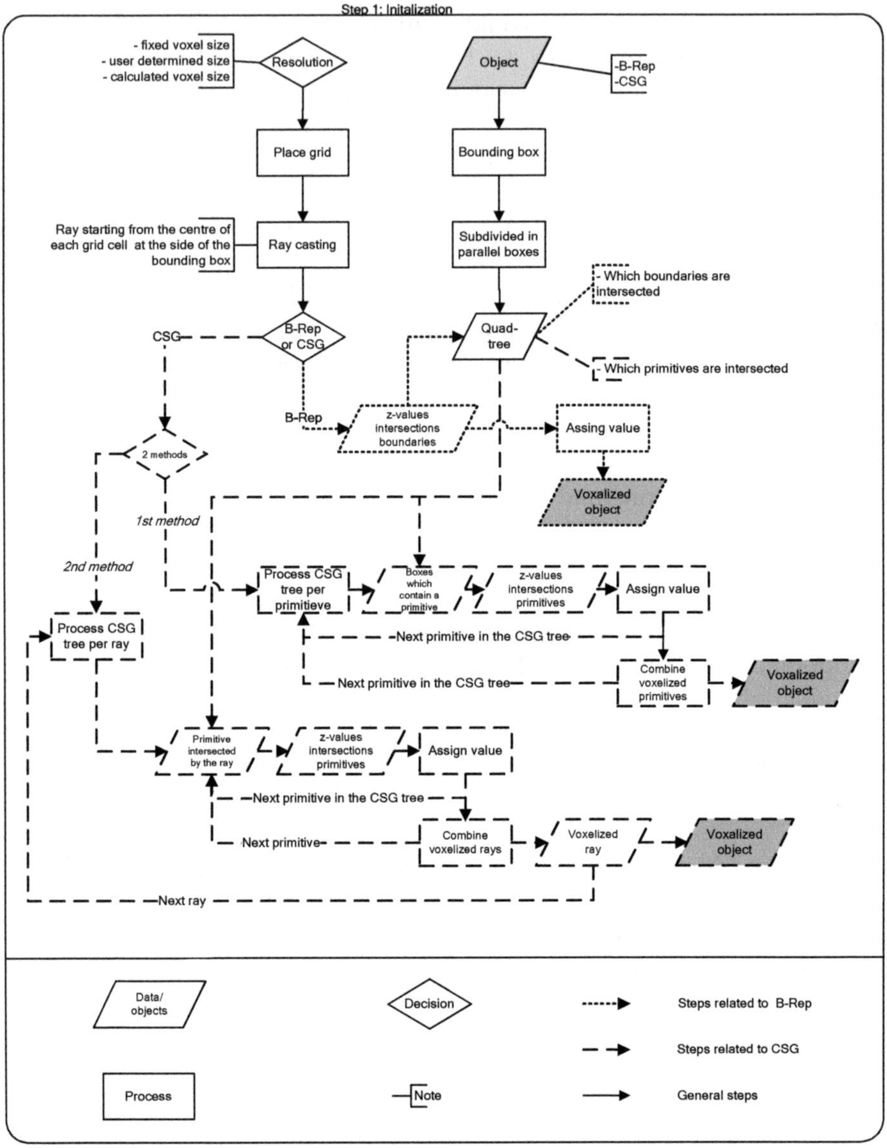

Fig. 3.: Scheme of the procedures followed in the first step

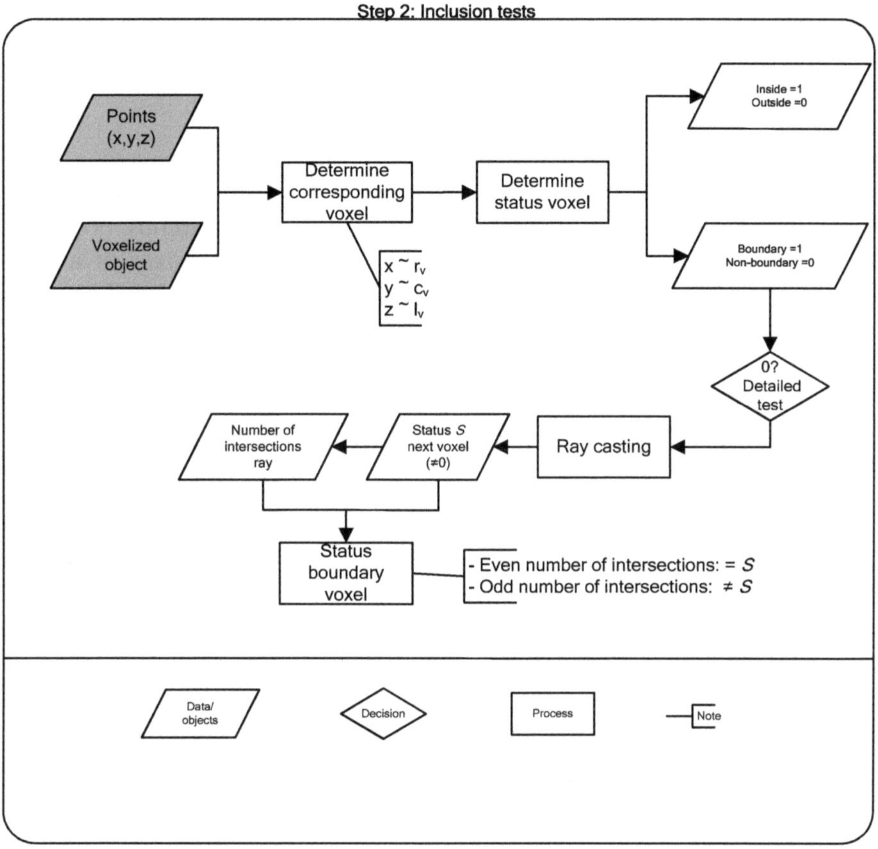

Fig. 4.: Scheme of the procedures followed in the second step

4 Conclusion and future work

The aim of this paper is to develop a method which can test very large datasets on a conventional desktop PC without any extra hardware requirements. An inclusion test with a preprocessing phase is considered the most appropriate to accomplish this. The method proposed in this paper is a uniform inclusion test which can handle two types of 3D object representations: the Boundary Representation and the Constructive Solid Geometry.

The proposed inclusion test is a 3D extension of the 2D Cell Based Containment Algorithm described by Zalik and Kolingerova (2001). The method is adapted such that it can handle both B-Rep and CSG object representations. The general principle of the method is the same for both types; there is only a slight difference in the algorithms for the two types during the voxelization phase.

The tests described above relate to only one polyhedron and a large number of points. The methods can easily be extended to multiple polyhedrons, each within its own bounding box or within overlapping bounding boxes. When multiple objects are present, the bounding boxes of all objects are relevant. A uniform, regular grid of voxels is then placed over the complete scene. Only the voxels which are located within these bounding boxes need to be considered during the voxelization procedure to derive their status (inside/outside). All the voxels located outside the bounding boxes cannot be located within any of the objects and thus receive by default the status outside.

Next, the algorithm processes each group of voxels located (partly) within a bounding box of a certain object. In order to facilitate further calculations, the size and location of the bounding box are aligned with the location of the bounding voxels around the object. In a following step, each individual object in voxelized separately. A special case emerges when several bounding boxes overlap. In this case, the results from the voxelization procedures, originating from the different objects, need to be combined at the locations where the overlap occurs. Only when the status of the cells is outside in all results, the cell is to be considered 'outside' the object and 'inside' otherwise.

Currently, the methods presented above are only described in a theoretical way. Their acceptability, in terms of the processing times, is studied with using complexity theory. The final goal in this research is to implement the methods described above, in order to be able to empirically test the exact values of the processing tasks. This allows verifying the obtained theoretical complexity functions.

References

DeMers MN (2000) Unit 4: Analysis: the heart of the GIS. In: DeMers MN, Fundamentals of Geographic Information Systems. John Wiley & Sons, Inc, New York, pp 181-395

Fang S, Chen H (2000) Hardware accelerated voxelization. Computers & Graphics, 24: 433-442

Feito F, Torres JC (1997) Inclusion test for general polyhedra. Computers & Graphics, 21(1): 23-30

Feito F, Torres JC, Urena A (1995) Orientation, simplicity and inclusion test for planar polygons. Computers & Graphics, 19(4): 595-600

Gombosi M, Zalik B (2005) Point-in-polygon tests for geometric buffers. Computers & Geosciences, 31: 1201-1212

Goodchild MF (1990) Simple Algorithms II – Polygons. In: Goodchild MF, Kemp KK (eds), NCGIA Core Curriculum in GIS. National Center for Geographic Information and Analysis, University of California, Santa Barbara CA

Haines E (1994) Point in Polygon Strategies. In: Heckbert P (ed), Graphics Gems IV. Academic Press, Boston, MA, pp 24-46

Hormann K, Agathos A (2001) The Point in Polygon Problem for Arbitrary Polygons. Computational Geometry, 20(3): 131-144

Huang C-W, Shih T-Y (1997) On the complexity of point-in-polygon algorithms. Computers & Geosciences, 23(1): 109-118

Li J, Wang W, Wu E (2007) Point-in-Polygon tests by convex decomposition. Computers & Graphics, 31: 636-648

Ogayar CJ, Segura RJ, Feito FR (2005) Point in solid strategies. Computers & Graphics, 29(4): 616-624

Preparata FP, Shamos MI (1985) Computational Geometry: an Introduction, 2nd edn, Springer, New York, pp 398

Stoter JE, Zlatanova S (2003) 3D GIS, where are we standing? ISPRS Joint Workshop on 'Spatial, Temporal and Multi-Dimensional Data Modelling and analysis', Québec, October

Taylor G (1994) Point in Polygon Test. Survey Review 32, 479-484

Thon S, Gesquière G, Raffin R (2004) A Low Cost Antialiased Space Filled Voxelization Of Polygonal Objects. GraphiCon 2004 proceedings, Moscou, pp 71-78

Walker RJ, Snoeyink J (1999) Practical point-in-polygon tests using CSG representations of polygons. In: Goodrich MT and McGeoch CC (eds) Algorithm Engineering and Experimentation (Proc. ALENEX '99). LNCS vol 1619, Springer-Verslag, pp 114-123

Wang W, Li J, Wu E (2005) 2D point-in-polygon test by classifying edges into layers. Computers & Graphics, 29: 427-439

Zalik B, Kolingerova I (2001) A cell-based point-in-polygon algorithm suitable for large sets of points. Computers & Geosciences, 27: 1135-1145

3D Volumetric Soft Geo-objects for Dynamic Urban Runoff Modeling

Izham Mohamad Yusoff[1], Muhamad Uznir Ujang[1] and Alias Abdul Rahman[1]

1 Department of Geoinformatics,
Faculty of Geoinformation Science and Engineering,
University Technology Malaysia,
81310 UTM Skudai, Johor Bahru, Malaysi
izham.fksg@gmail.com, {mduznir,alias}@utm.my

Abstract. Complexities of streamflow structures, drainage systems, sewers and urban landscape render difficulties for urban flood mitigation and rehabilitation planning towards reducing urban flood disaster. GIS has been useful for the investigation of urban runoff spatial patterns. However, the inability of GIS to realistically explore the momentum and continuity aspects of overland flow and urban runoff process has been proven to be the shortcomings of the system. This leads to the requirement of an extended dimension representing dynamic soft geometrical volumes of urban stormwater runoff. The inclusion of Volumetric Soft Geo-objects (VSG) offers a comprehensive and practical 3D dynamic modeling of urban runoff volume. The VSG are driven based on Kinematic Wave Routing and Green-Ampt method for modeling open channel flow and overland flow volume respectively. Basin model, sub-basin, reach and junction elements are extracted from Digital Elevation Model (DEM) with 5 meter resolution using the HEC-GeoHMS program and the HEC-HMS model. The modeled results of discharge volume gave an R^2 of 0.88 and a Nash–Sutcliffe coefficient of 0.82, indicating that VSG provides end-users the prospect to envision complex information of flood disaster management.

Keywords: 3D GIS, Continuity, Green-Ampt Method, Kinematic Wave Routing, Momentum, Visualization and VSG.

1. Introduction

River basins in Malaysia are now under intense pressure from urban, industrial, and infrastructural development which increases the percentage of impervious area in a catchment. The problems become even more aggravated by frequent intense rainfalls, the physiological nature of basins and poor urban

T. Neutens, P. De Maeyer (eds.), *Developments in 3D Geo-Information Sciences*,
Lecture Notes in Geoinformation and Cartography, DOI 10.1007/978-3-642-04791-6_11,

services. The Department of Irrigation and Drainage, Malaysia (DID) estimates an annual cost of damage of RM2.5 billion due to flood events. Various urban runoff models and methods have been developed and integrated with GIS to model the loss, transformation and baseflow rate within a basin such as SCS Technical Release (TR-55), Storage Treatment Overflow Runoff Model (STORM), Distributed Routing Rainfall-Runoff Model (DR3M), MIKE-SHE, AGWA, Soil and Water Assessment Tools (SWAT), Kinematic Runoff and Erosion Model – KINEROS and Storm Water Management Model – SWMM. These models can be used to access the runoff rate, analyze an existing network of interconnected stormwater management facilities, and design new components such as underground storm sewers, detention ponds, ditches, channels and street-side curbs of an existing system. However, the models merely yielded limited success. The urban runoff has become complex due to the complicated structures and distributions of urban components. This leads to the need of enhancement of visualization techniques. Since the volume, rate, and velocity of runoff from a particular rain event will depend on the characteristics of the surfaces, changes to these surfaces can significantly change the resultant overland flow volume, runoff rate and velocity (Wang *et al.*, 2007).

Rapid development in computer graphics has provided 2D GIS users with a new dimension of visualization (3D) for specific natural disaster applications such as the modelling of urban floods, landslides and coastal erosions. The issues pertaining to dynamism has encouraged 3D GIS practitioners to represent changes of 3D data models, to index 3D dynamic spatial objects and to analyze the influence of event based parameters. The parameters include precipitation input, soil infiltration capacity, groundwater flow and the increase of stormwater discharge within streamflows. In this paper, the authors used the term dynamic to refer to the momentum and continuity of overland flow as the reference of the model computation on the plane surfaces. The computation is based on the equation of open channel flow to represent the continuity in the sub-basin channels.

At present, the large variation and behavior of drainage basin render difficulties of producing general relationships of the physical soft-geometrical features (i.e. stormwater). Lin *et al.* (2008) applied the Virtual Geographic Environment (VGE) concept based on GIS and Virtual Reality technologies for dynamic modelling of atmospheric pollution dispersion in city clusters. This is carried out by maintaining an Octree to construct a view dependent-representation of regular volume data, which resides under hybrid geometry types. Beni *et al.*, (2007) proposed a dynamic 3D data structure based on 3D Delaunay tetrahedralization that deals with objects and field representation of volume and space at the same time. This provides an on-the-fly interactive topological network (grid) for numerical modeling. These research approaches can be classified under rigid geo-objects. However, soft geo-objects (i.e. streams, mudflows) are less represented. Shen *et al.* (2006) represented a 3D modeling of soft geo-objects representing overland flow based on the GIS flow element (FE) concept. However, the aspect of volumetric dynamic flow is not visualized. A previous study by (Izham *et al.*, 2008) revealed the influence of georeference and plane fitting techniques on 3D spatial objects, and preliminary implementation of VSG representing dynamism of infiltrated stormwater and overland flow.

This paper enhances the details of VSG modelling. Emphasis is given to the momentum and continuity of urban runoff process, verification of modeled VSG results and 3D visualization techniques, using soft geo-objects approach. The VSG method in this study is rendered to visualize urban runoff mechanisms and enable GIS practitioners to access hydrological-soil impacts, complex infiltration, channel flow and urban runoff information. Modeling is performed within basin boundaries driven by the Kinematic Wave Routing method and Green-Ampt infiltration equation. The computation of open channel flow and overland flow is carried out by using the methods aforementioned respectively, via the metaball approach. The concepts of 3D dynamic urban runoff modeling and VSG are explained in section 2. The experiment of determining VSG modeling for open channel flow and overland flow is highlighted in Section 3. Urban flood areas and 3D dynamic urban runoff modeling results are explained in Section 4. Section 5 concludes the findings.

2. VSG for Urban Runoff Modeling

2.1. 3D GIS, Momentum and Continuity

Current 3D hydrologic methods are insufficient for the representation of fluids as soft geo-objects. Urban runoff occurs from the portion of rain, irrigation water or wastewater that does not infiltrate the soil, reach streams and channels, by travelling over the surface of the soil (Bedient and Huber, 2002; Ward and Trimble, 2004). Significant contributions of 3D GIS for urban runoff modelling can be determined by its ability to visualize and integrate the spatial connection and the influences between layers. For instance, the momentums of overland flow and open channel flow requires soils to be saturated, exceed the depression storage and flow towards low elevation areas continually through soil layers and channelized stormwater network. Incorporating and linking the urban runoff process within spatial layers of precipitation, soil, DEMs and channel network could be coupled successfully with the use of 3D GIS approach.

An algorithm called the Flood Region Spreading Algorithm (FRSA) developed by Wang *et al.* (2007) explores the pseudo-intensity profile for potential flood spreading. The flood region search and merge process are conducted by a weighted cell region adjacency graph. The similarity of the measure associated with two neighboring cells is found by calculating the mean intensity values of all vertices associated for each cell, and the boundary between the two. The down cell will be flooded if the flooded cells (up cell) mean value is greater than the down cell's mean value, and greater than the minimum value of the common boundary. A connected cell is identified as part of a flooded cell of the flood region if its elevation level is under a dam threshold of a neighboring flooded cell. The process is conducted in all eight neighboring directions around the cell assessed. Flooded cells are merged to become a large cell. The configured flood region consists of all flooded cells connected together and used as a flooding field.

The data structure proposed by Beni *et al.* (2007) has the ability to generate an optimal network (grid) for numerical modeling. In this network, local updates of topology are possible after any change (event). The network allows fluid flow such as flood or forest fire to be efficiently modeled within GIS if the Free-Lagrange method is used as the numerical model solution.

In this study, the spread of urban runoff, spatial connection and influences of urban components are visualized and enhanced through VSG, driven by the infiltrability of soil surface computed using the Kinematic Wave Routing and Green-Ampt method. This enables dynamic visualization of runoff direction by merging soft geo-objects of overland flow and open channel flow. Moreover, the method allows interpretation of routing through the use of different urban flood management strategies combined with climate change patterns. The combination of visualization techniques and modeling model provided by 3D GIS allows querying of the data model, integration with other datasets using a common format, and transfer between the modules used for visualization. By developing urban runoff modeling, specific analysis can potentially provide a replicable, rational and transparent method to explore the complex processes of the overland flow and open channel flow process within a structured framework. Therefore, modelling VSG momentum and continuity of urban runoff process involve simplification of a complex environment in order to improve understanding in a systematic manner.

2.2. Mathematical Kinematic Wave Routing and Green-Ampt Computation

The Kinematic Wave routing method represents a sub-basin (a) as a wide open channel (b) with inflow to the channel equal to the excess precipitation as illustrated in Figure 2.1 (Ward and Trimble, 2004; Feldman, 2000). It solves the equations that model unsteady shallow water flow in an open channel to compute the watershed runoff hydrograph. The method represents the sub-basin as two plane surfaces over which water runs until it reaches the channel. The water then flows down the channel to the outlet. At a cross section, the system would resemble an open book, with the water running parallel to the text on the page (down the shaded planes) and then into the channel that follows the book's center binding. The kinematic wave routing model represents momentum of direct runoff on the plane surfaces and flows continually in the sub-basin channels.

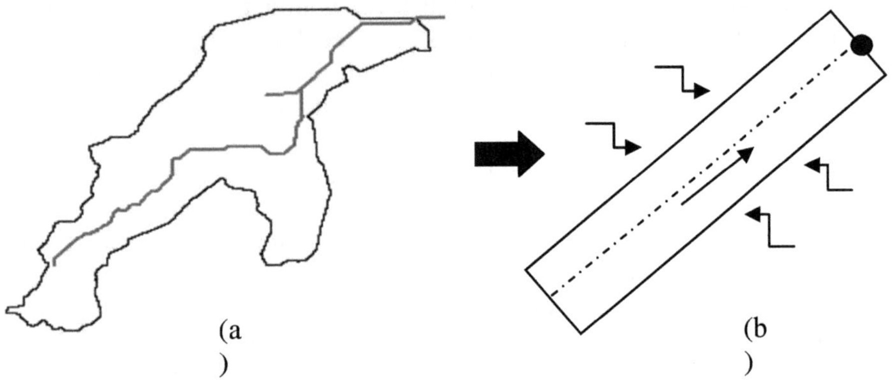

Fig. 2.1. Simple watershed with kinematic-wave model representation.

Direct runoff and open channel flow is computed based on the concept of overland flow mentioned by Chow (1959) and Chaudhry (2008). The reference of the overland model computation depends on the equation of open channel flow, which is the momentum equation and the continuity equation.

In one dimension, the momentum equation is :

$$Sf = So - \frac{\partial y}{\partial x} - \frac{v}{g}\frac{\partial v}{\partial x} - \frac{1}{g}\frac{\partial v}{\partial t} \quad \textbf{(1)}$$

where Sf = friction slope; So = bottom slope; V = velocity; $\frac{\partial y}{\partial x}$ = pressure gradient;
$\frac{v}{g}\frac{\partial v}{\partial x}$ = convective acceleration and $\frac{1}{g}\frac{\partial v}{\partial t}$ = local acceleration. The energy gradient can be estimated using Manning's equation (2).

$$Q = \frac{CR^{2/3}Sf^{1/2}}{N}A \tag{2}$$

where Q = flow, R = hydraulic radius, A = cross-sectional area and N = a resistance factor that depends on the cover of the planes. Equation (2) can be simplified to :

$$Q = \alpha A^m \tag{3}$$

where α and m are parameters related to flow geometry and surface roughness. The second critical continuity equation is

$$A\frac{\partial v}{\partial x} + VB\frac{\partial v}{\partial x} + B\frac{\partial y}{\partial t} = q \tag{4}$$

where B = water surface width; q = lateral inflow per unit length of channel; A = prism storage and VB = wedge storage and $B\frac{\partial y}{\partial t}$ = rate of rise. The lateral inflow represents the precipitation excess, computed as the difference in precipitation losses. With the simplification appropriate for shallow flow over a plane, the continuity equation reduces to:

$$\frac{\partial A}{\partial t} + \frac{\partial Q}{\partial x} = q \tag{5}$$

The combination of equations (3) and (4) lead to Equation (6), which is a kinematic-wave approximation of the equations of motion.

$$\frac{\partial A}{\partial t} + \acute{a}mA^{(m-1)}\frac{\partial A}{\partial x} = q \tag{6}$$

The Green-Ampt infiltration model is a physical model which relates the rate of infiltration to measurable soil properties such as the porosity, hydraulic conductivity and the moisture content of a particular soil based on simplified solutions to the Richards equation. The equation for infiltration under constant rainfall based on Darcy's law assumes a capillary tube analogy for flow in a porous soil as follows:

$$f = K(H_o + S_w + L)/L \tag{7}$$

where K is the hydraulic conductivity of the transmission zone, H_o is the depth of flow ponded at the surface, S_w is the effective suction at the wetting front, and L is the depth from the surface to the wetting front. The method assumes piston flow (water moving down as a front with no mixing) and a distinct wetting front between the infiltration zone and soil at the initial water content. Smemoe (1999) stated the basic Green and Ampt equation for calculating soil infiltration rate as follows :

$$f = K_s\,(1 + [\Psi\theta / F]) \quad \mathbf{(8)}$$

where Ks is the saturated hydraulic conductivity, Ψ is average capillary suction in the wetted zone, θ is soil moisture deficit (dimensionless), equal to the effective soil porosity times the difference in final and initial volumetric soil saturations and F = depth of rainfall that has infiltrated into the soil since the beginning of rainfall.

Specific prediction capabilities of the Kinematic Wave routing and Green-Ampt method would be vital to integrate dynamic VSG modeling and visualization to compute the amount of open channel flow volume. The capability of GIS techniques to analyze infiltration and open channel flow as mentioned by Bedient and Huber (2002), Ward and Trimble (2004), Brutsaert (2005) and Shen *et al.* (2006) are realized by producing VSG features driven by Kinematic Wave routing method to visualize urban runoff areas.

3. 3D VSG Urban Runoff Experiment

3.1. Study Area

The Sungai Pinang basin lies between the Latitude of 5° 21' 32" to 5° 26' 48" North and Longitude from 100° 14' 26" to 100° 19' 42" East. The Sungai Pinang is the main river system in the Penang Island with an approximated basin size of 51 km^2 , comprising mainly the urban areas of "George Town", "Air Hitam" and "Paya Terubong" as depicted in Figure 3.1. Penang Island is located at the West Coast of Peninsular Malaysia and experiences convective storms generated by the inter monsoon seasons (Sumatra wind system) in April/May and October/November. The South-West Monsoon (normally from May to September) produces less rain in the West Coast of the Peninsular whilst the North-East Monsoon, from November to March, carries longer and heavier rains to the East Coast of the Peninsular, North Sabah, and inland Sarawak (USMM, 2000). Penang Island is characterized by the equatorial climate, which is warm and humid throughout the year and has an average annual rainfall of more than 2477 mm; with the lowest monthly average around 60 mm for February and the highest around 210 mm for August and October. The Sungai Pinang basin is a highly developed area comprising more than 40 percent of urban areas in the Penang Island. The Sungai Pinang basin has been selected to determine saturation and overland flow volume process due to the continuity of development that affects the physical characteristics of land use and soils. The effects include the degradation of water quality and increase of water quantity in the entire basin. Moreover, flash flood and water pollution are the main problems that occur in highly urbanized areas such as "Georgetown", "Jelutong" and "Air Hitam".

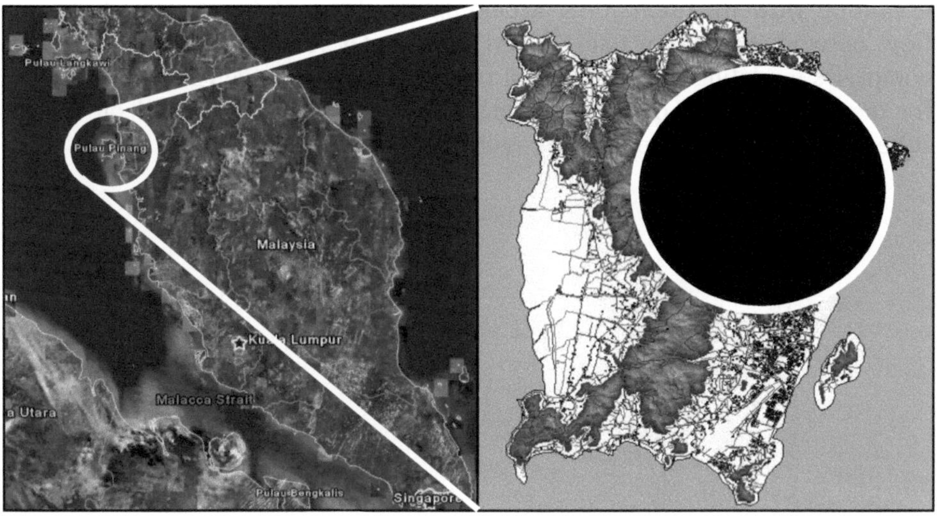

Fig. 3.1. Location of Sungai Pinang basin.

The procedure for linking GIS with flow routing parameters and 3D dynamic modeling of urban runoff process involves the following steps: (1) acquisition and development of GIS map data layers of Sungai Pinang basin in Cassini-Soldner projection plane; (2) pre-processing of Kinematic Wave Routing and Green-Ampt model data input; parameters and computation results within the HEC-GeoHMS and HEC-HMS models and (3) post-processing of all dynamic flow volume components resulted from 3D VSG modeling for displaying overland flow, open channel flow volume and urban flood areas. The Green-Ampt and Kinematic Wave Routing parameters are linked into PC-based ArcView GIS package and commercial 3D modelling software to store and display dynamic VSGs for urban runoff modelling.

3.2. Determining Potential Overland Flow, Open Channel Flow and Urban Runoff Areas

Analysis is performed in two phases as illustrated in Figure 3.2. The first phase deals with creating the basin and meteorological elements to be incorporated in the HEC-HMS model by using the Geospatial Hydrologic Modelling (HEC-GeoHMS) program within ESRI's ArcView GIS software. The elements are obtained through hydrologic modelling by filling sinks, and determining flow direction, flow accumulation, basin delineation and streamflow network extraction using DEM of 5 meter resolution. The next phase comprises modeling of precipitation-runoff processes in the HEC-HMS model for the identification of urban drainage and overland flow based on the resulted hydrograph as depicted in Figure 3.3.

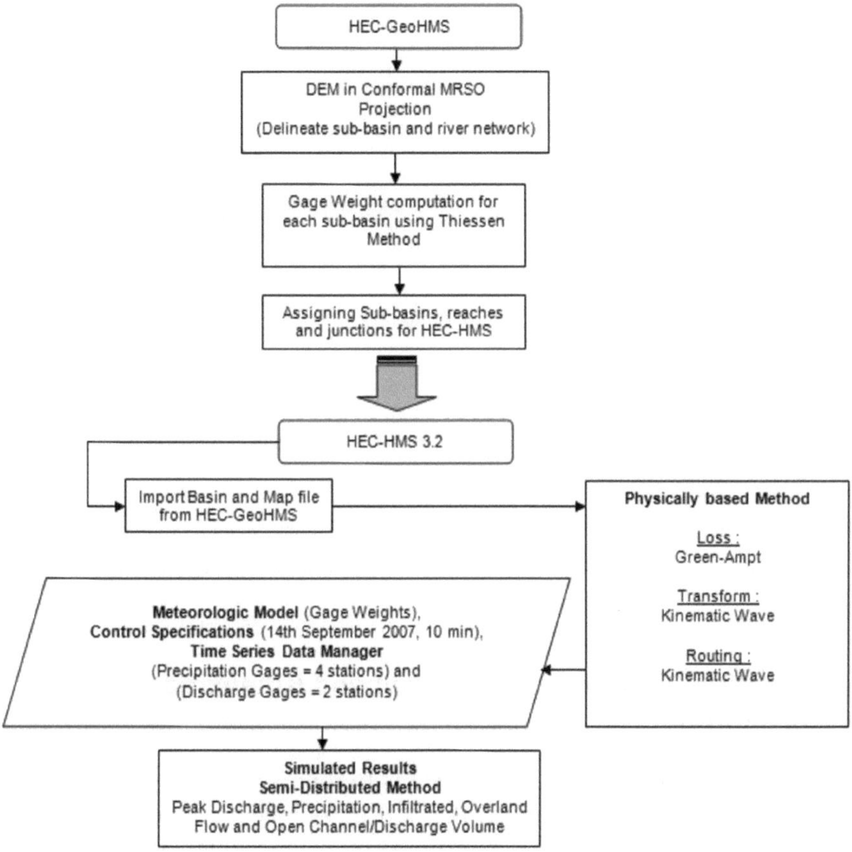

Fig. 3.2. Schematic diagrams for determining overland flow, open channel volume and flooded area.

To estimate the open channel flow using the kinematic wave routing method, each sub-basin is described as a set of elements that include overland flow planes, collector and sub-collector channels; and the main channel. The overland flow planes consist of information of its typical length, representative slope, overland flow roughness coefficient, area represented by plane and loss model parameters. The collector and sub-collector channels require inputs of area drained by channel, representative channel length, and channel shape, dimensions of channel cross section, channel slope and representative of Manning's roughness coefficient. The main channel comprises information of channel length, channel shape, dimensions of cross section, channel slope, representative Manning's roughness coefficient and identification of inflow hydrograph.

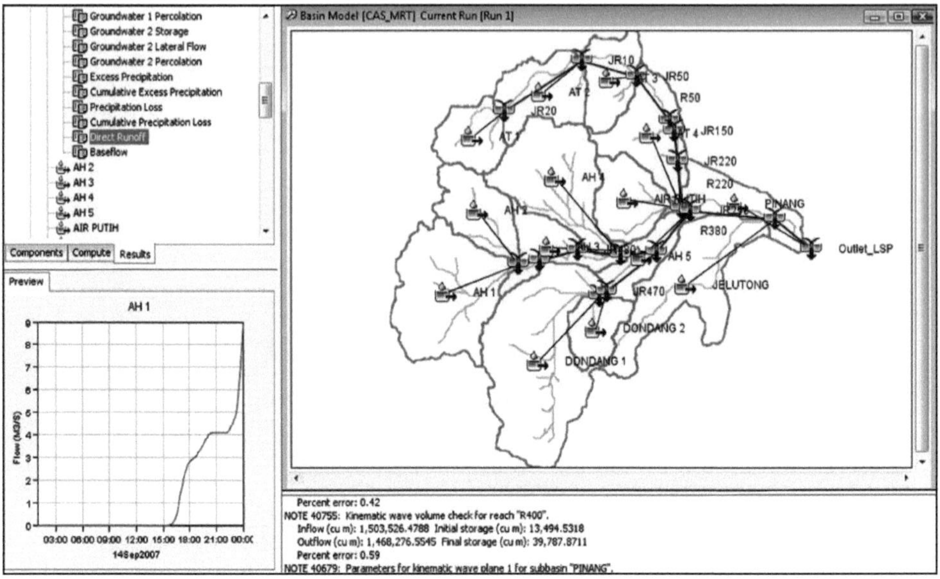

Fig. 3.3. Delineated sub-basin boundaries, reaches and junctions from HEC-HMS hydrologic model.

3.3 Integrating VSG with Green-Ampt and Kinematic Wave Routing Method

The key objective of developing VSG is to represent a temporal-based continuous spreading of urban runoff, the soft geo-objects connection and influences of urban components (i.e. buildings, roads and other 3D hardscape objects). The distribution of VSG is rendered as a volume and reflects the dynamism by merging volumes of runoff, changes of velocity and directions. The dynamic VSG flow is identified by using eight neighboring pixels (D8) introduced by Tarboton (1997) to determine flow direction and velocity based on slope gradient.

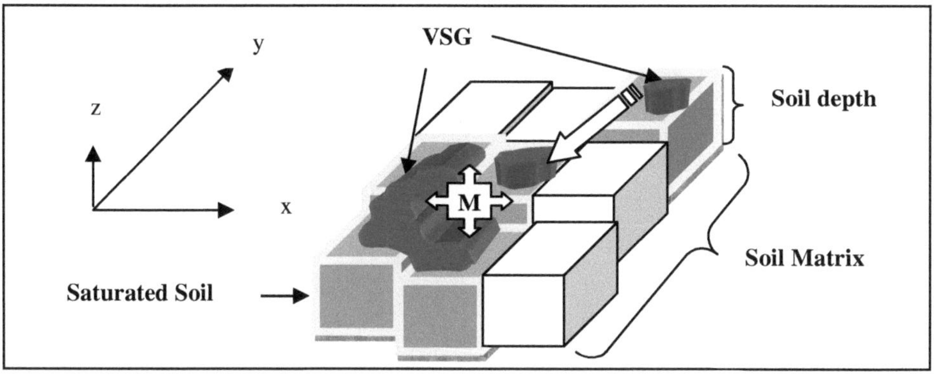

Fig. 3.4. VSG generated due to saturated soil layer, flows towards low elevation and merges (M).

The generation of VSG is based on top soil infiltrability. Continuous precipitation input on saturated soil causes the stormwater exceeds the soil infiltrability using equation (7) and (8); and distributes VSGs, which carries values of overland flow (precipitation minus infiltration). The shapes of VSGs are proportional with the topographic surface of the basin. The next modeling process consists of open channel flow driven by Kinematic Wave Routing method. VSGs are integrated with the method and shapes according to channel designation as quoted in equation (6). The 3D dynamic modeling involves the merging of VSGs (depicted in Figure 3.4) and visualization of temporal-based overland flow, open channel flow and flow direction on top soils. As urban stormwater runoff decreases, VSGs are omitted due to re-infiltration of top soils. Additional textures on VSGs would deliver information such as the high and low infiltrated urban stormwater runoff, velocity of VSGs, overland flow and open channel flow within basin area.

3.4. Computation and 3D Dynamic VSG Visualization of Open Channel Flow, Overland Flow and Urban Runoff Areas

Simulations of VSG are performed based on modeled overland flow using equation (7) and (8), while open channel flows are modeled by referring to equation (6). The total of open channel flow and overland flow within the Sungai Pinang basin are computed by subtracting precipitation volume with the infiltrated rainfall volume using the rainfall data recorded on 14th of September, 2007 with 10 minutes interval. Figure 3.5 shows the flow diagram determining 3D dynamic VSG for urban runoff modeling represented by a fine cylinder.

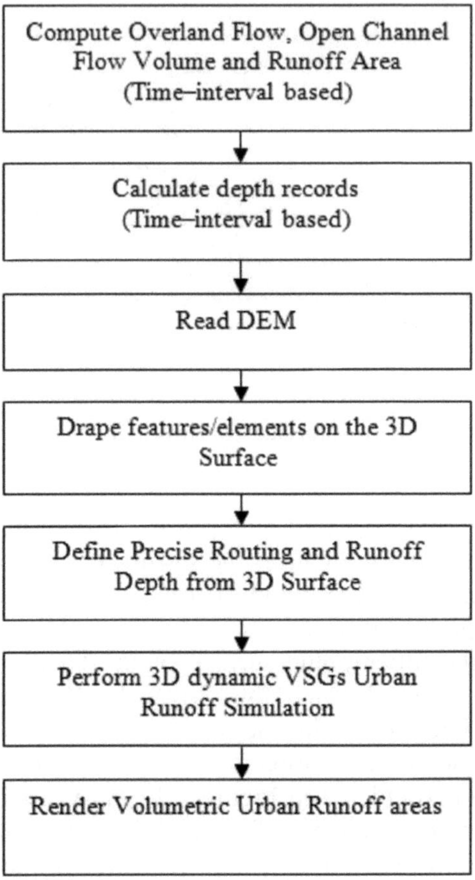

Fig. 3.5. Flow diagram for rendering 3D dynamic urban runoff VSG modeling.

4. Results and Discussion

4.1. Potential Urban Runoff Areas

The experiment of determining potential urban runoff area is illustrated in Figure 4.1. Approximately 11.59 km^2 of urban runoff areas are identified. Most of the urban runoff coverage lies in areas of "Georgetown", "Paya Terubong", "Air Hitam", "Air Terjun River", "Kebun Bunga", "Green Lane" and partly in "Gelugur" and "Jelutong". The location of urban runoff area lies on the sub humid to humid regions, which are the major controls on the various urban runoff processes based on meteorological factors and physical characteristics as stated by USGS (2005).

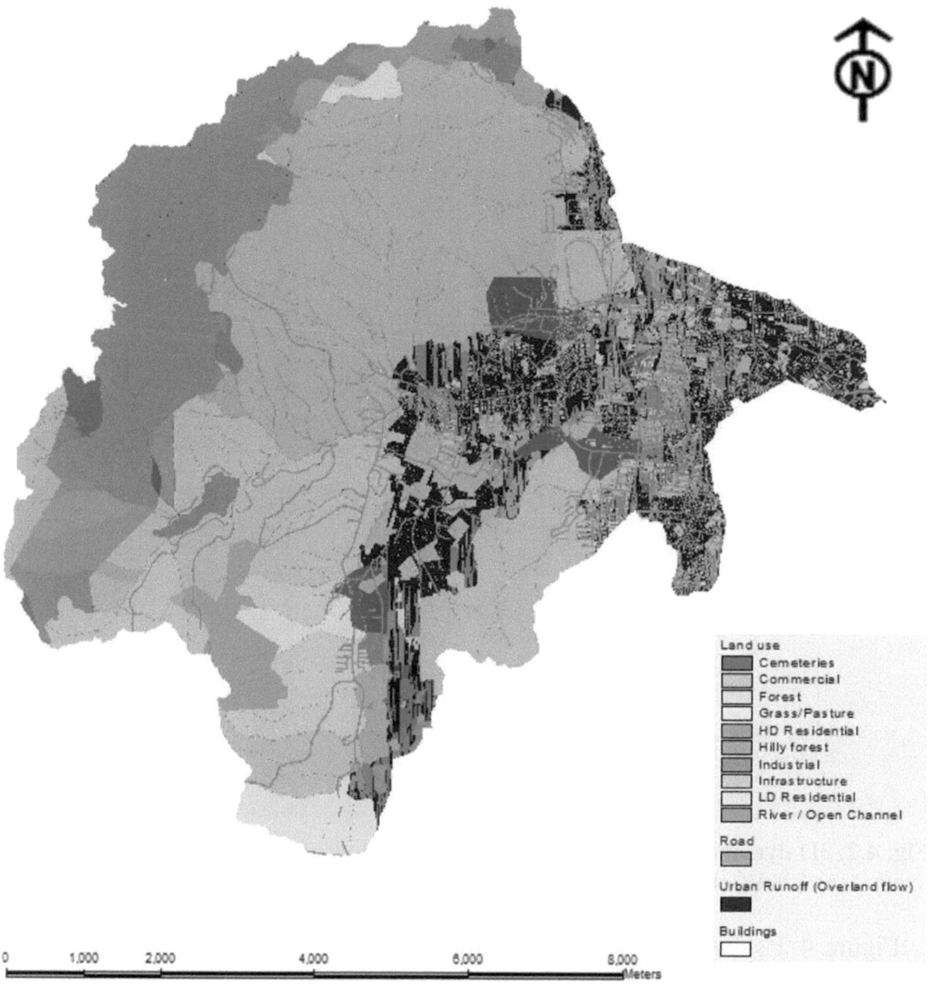

Fig. 4.1. Potential urban runoff area consists of overland flow and open channel flow within Sungai Pinang basin based on 14th September 2007 rainfall data.

4.2. 3D Dynamic VSG Modeling of Overland Flow, Open Channel Flow Volume and Urban Runoff Areas

Approximately 5,114,100 m^3 of precipitation volume were recorded within the Sungai Pinang basin. The estimated volume of rainfall infiltrating soil is 1,197,000 m^3. The total of overland flow and open channel flow volumes modeled by VSG are estimated at 968,400 m^3 and 1,718,000 m^3 respectively. The results obtained are illustrated in Figure 4.2 and 4.3. The full summary of the analyzed overland flow and open channel flow volume using Green-Ampt and Kinematic Wave Routing methods for each sub-basin in the Sungai Pinang basin is illustrated in Table 4.1.

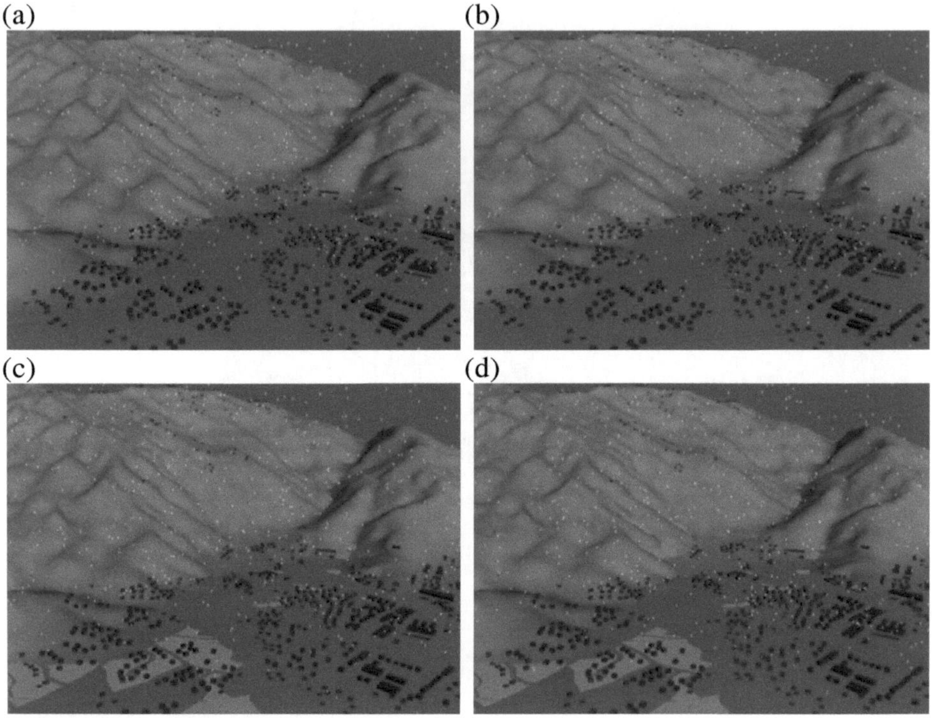

Fig. 4.2. 3D dynamic VSG modeling for overland flow and open channel flow visualized at (a) 1 hour, (b) 6 hours, (c) 12 hours and (d) 18 hours rainfall data.

Figure 4.2 shows the momentum and continuity of 3D VSG to determine the open channel flow and overland flow volumes, projected under the Cassini-Soldner plane. Continuous input from precipitation increases the height and coverage of the overland flow and open channel flow volumes, mainly on downslopes and flat areas. The modeled VSG is proportional to the Kinematic Wave Routing method, Green-Ampt method and physical characteristics of the equidistant Cassini-Soldner projection plane. The outflow from all VSGs are collected and merged into ordinary rendered settings, representing the overland flow volume, open channel flow volume and direction, flow discharge and flooded areas.

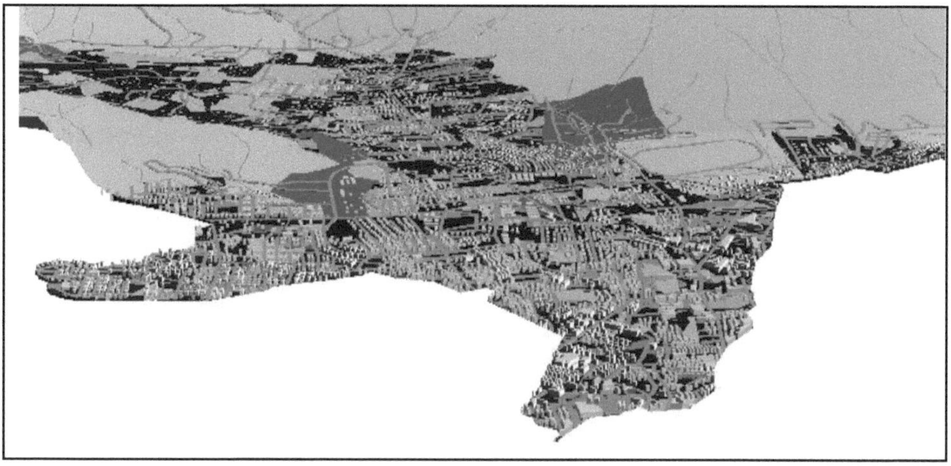

Fig. 4.3. Flooded areas visualized using VSGs at the outlet of Sungai Pinang basin.

Figure 4.3 illustrates the flooded areas in 3D environment (shaded with black) which lie in sub-basins of "Air Hitam 3", "Air Hitam 4", "Air Hitam 5", "Air Putih", "Jelutong" and "Sungai Pinang". The VSG visualizes flooded areas based on impervious area coverage, overland flow and channel flow that spill out from the existing drainage and streamflow pattern. Construction of shop lots, apartments and widened road network increases the land cover with impervious areas. This is the main factor contributing to urban flooding, thus indicating that the existing rivers and drainage systems lack the capability of shifting runoff volumes from highly urbanized areas. The modeled results are then verified and compared with observed discharge volume as shown in Figure 4.4.

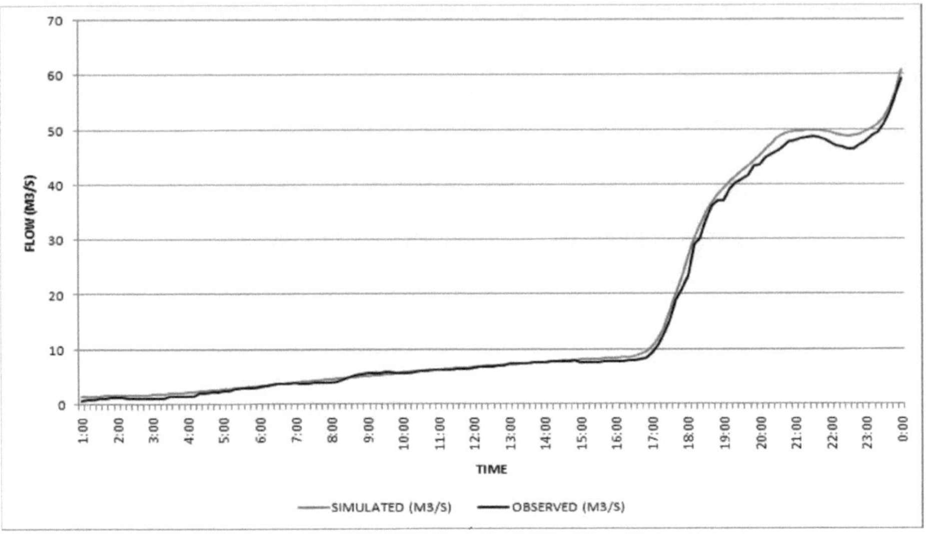

Fig. 4.4. Comparison of Modeled Discharge Volume with Observed Discharge Volume

The modeled urban runoff discharge volume corresponds well to the observed measurements. Figure 4.5 illustrates the R^2 error computed with observed streamflow at the outlet of the basin.

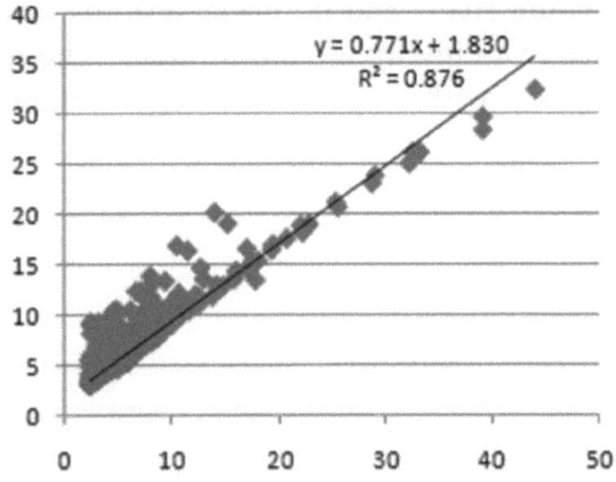

Fig. 4.5. R^2 of Modeled Discharge Volume.

The comparison of modeled and observed discharge volume using 10 minutes interval gives an R^2 of 0.88 and Nash–Sutcliffe coefficient of 0.82 as the performance indicator of hydrologic process. The modeled results indicate that the 3D dynamic VSG urban runoff modeling indeed provides a valuable step for end-users visualizing complex volume information in 3D environment. The visualization includes flow routing, floodplain areas, affected buildings and land properties.

Table 4.1. Modeled VSG urban runoff process results using Kinematic Wave Routing and Green-Ampt method.

Sub-basin	Peak Discharge (M^3/s)	Precipitation Volume (1000 M^3)	Infiltrated Volume (1000 M^3)	Overland Flow Volume (1000 M^3)	Open Channel / Discharge Volume (1000 M^3)
Air Hitam 1	9.9	640.6	241.5	100.2	150.0
Air Hitam 2	6.9	271.8	83.8	83.4	114.2
Air Hitam 3	5.5	228.4	29.1	85.3	116.3
Air Hitam 4	12.9	516.9	71.8	128.6	200.1
Air Hitam 5	1.5	182.7	18.2	12.3	31.1
Air Putih	4.0	266.4	26.5	42.4	67.4
Air Terjun 1	4.2	214.0	71.5	61.2	89.0
Air Terjun 2	7.2	309.9	90.1	81.3	115.8
Air Terjun 3	7.7	243.5	31.0	125.3	158.0
Air Terjun 4	9.1	360.8	50.1	99.9	159.0
Dondang 1	5.8	877.6	306.7	40.0	174.4
Dondang	4.6	184.6	40.9	71.3	117.9
Jelutong	5.1	549.3	101.2	34.4	151.6
Pinang River	1.4	267.6	34.6	2.8	73.2
Total	-	5114.1	1197.0	968.4	1718.0

The differential of modeled and observed discharge volume is proportional to the DEM, the borders of rainfall, land use and soil coverage. The previous research carried out by Izham *et al.* (2008) emphasizes the importance of projection plane for preserving adjacent spatial objects, which directly connects to the land surface. Hence, the modeled results from VSG could be affected due to the different characteristics of shape, area, distance and direction of each 3D spatial object on the basin surface. Changes to these surfaces can significantly change the resultant overland flow volume, runoff rate and velocity as mentioned by Wang *et al.* (2007). The 3D VSG modeling performed however does not take the water balance equation into account. The equation includes evapotranspiration losses, percolation, return flow, groundwater flow, shallow and subsurface flow, which could potentially result with dissimilar computation of modeled and observed discharge volume.

5. Concluding Remarks

This paper discusses the definition, mathematical expression and representation of 3D dynamic modeling of urban runoff process. The authors employed the VSG approach to visualize the momentum and continuity of overland flow and open channel flow of the Sungai Pinang basin. This research is carried out by incorporating volumetric elements of previous studies on 3D soft geo-objects modeling by Shen *et al.* (2006). VSG is driven by the Kinematic Wave Routing and Green-Ampt method. A practical approach needs to be employed prior to any urban runoff modelling and disaster management within a 3D environment. Determining the 3D GIS properties (i.e. 3D topological aspects, 3D spatial indexing, 3D generalization) that can be utilized to represent hydrologic properties of a river basin is one such approach.

References

1. Bedient, P. B., & Huber, W. C. (2002). Hydrology and Floodplain Analysis. (3rd ed.) New Jersey: Prentice Hall, (Chapter 2, 5, 7 and 10).
2. Beni, L. H., Mostafavi, M. A., Pouliot, J. (2007). 3D Dynamic Simulation within GIS in Support of Disaster Management. Geomatics Solutions for Disaster Management, Lecture Notes in Geoinformation and Cartography. Springer. 165-184.
3. Brutsaert, W. (2005). Hydrology : An Introduction. New York: Cambridge University Press, (Chapter 11).
4. Chaudhry, H.C. (2008). Open Channel Flow. (2nd ed.) New Jersey: Prentice Hall, (Chapter 2).
5. Chow, V.T. (1959). Open Channel Flow. New York: McGraw-Hill, (Appendix A).
6. Feldman, R. D. (2000). Hydrologic Engineering Center – Hydrologic Modelling System Technical Reference Manual. California: US Army Corps of Engineers, (Chapter 8).
7. Izham, M. Y., Muhamad, U. U., Alias A. R. (2008). 3D Dynamic Simulation and Visualization for Infiltration Excess Overland Flow Modeling. 3D Geoinformation Sciences, Lecture Notes in Geoinformation and Cartography. Springer. 413-430.
8. Lin, H., Zhu, J., Xu, B., Lin, W., Hu, Y. (2008). A Virtual Geographic Environment for a Simulation of Air Pollution Dispersion in the Pearl River Delta (PRD) Region. 3D Geoinformation Sciences, Lecture Notes in Geoinformation and Cartography. Springer. 3-14.
9. USMM. (2000). Urban Stormwater Management for Malaysia. Vol. 1 – 20. Department of Irrigation and Drainage. National Printing Malaysia Limited, Malaysia.

10. Shen, D. Y., Takara, K., Tachikawa, Y., Liu, Y. L. (2006). 3D Simulation of Soft Geo-objects. International Journal of Geographical Information Science, (20), 261-271.
11. Smemoe, C. M. (1999). The Spatial Computation of Sub-basin Green and Ampt Parameters. Unpublished.
12. Tarboton, D. G. (1997). A New Method for the Determination of Flow Directions and Upslope Areas in Grid Digital Elevation Models. Water Resources Research (33), 309-319.
13. USGS. (2005). United States Geological Survey : Earth's water – Runoff. http://ga.water.usgs.gov/edu/runoff.html
14. Wang, C., Wan, T. R., Palmer, I. J. (2007). A Real-time Dynamic Simulation Scheme for Large Scale Flood Hazard using 3D Real World Data. 11th International Conference Information Visualization (IV'07), IEEE Computer Society Xplore. 607-612.
15. Ward, A.D., & Trimble, S. W. (2004). Environmental Hydrology. (2nd ed.), Washington: Lewis Publishers, (Chapter 1 and 5).

MIX
Papier aus verantwortungsvollen Quellen
Paper from responsible sources
FSC® C105338

If you have any concerns about our products,
you can contact us on
ProductSafety@springernature.com

In case Publisher is established outside the EU,
the EU authorized representative is:
Springer Nature Customer Service Center GmbH
Europaplatz 3, 69115 Heidelberg, Germany

Printed by Libri Plureos GmbH
in Hamburg, Germany